Mobile Technologies in Libraries

LIBRARY INFORMATION TECHNOLOGY ASSOCIATION (LITA) GUIDES

Marta Mestrovic Deyrup, PhD
Acquisitions Editor, Library Information and Technology Association, a division of the American Library Association

The Library Information Technology Association (LITA) Guides provide information and guidance on topics related to cutting edge technology for library and IT specialists.

Written by top professionals in the field of technology, the guides are sought after by librarians wishing to learn a new skill or to become current in today's best practices.

Each book in the series has been overseen editorially since conception by LITA and reviewed by LITA members with special expertise in the specialty area of the book.

Established in 1966, the Library and Information Technology Association is the division of the American Library Association (ALA) that provides its members and the library and information science community as a whole with a forum for discussion, an environment for learning, and a program for actions on the design, development, and implementation of automated and technological systems in the library and information science field.

Approximately 25 LITA Guides were published by Neal-Schuman and ALA between 2007 and 2015. Rowman & Littlefield took over publication of the series beginning in late 2015. Books in the series published by Rowman & Littlefield are

Digitizing Flat Media: Principles and Practices
The Librarian's Introduction to Programming Languages
Library Service Design: A LITA Guide to Holistic Assessment, Insight, and Improvement
Data Visualization: A Guide to Visual Storytelling for Librarians
Mobile Technologies in Libraries: A LITA Guide

Mobile Technologies in Libraries

A LITA Guide

Ben Rawlins

ROWMAN & LITTLEFIELD
Lanham • Boulder • New York • London

Published by Rowman & Littlefield
A wholly owned subsidiary of The Rowman & Littlefield Publishing Group, Inc.
4501 Forbes Boulevard, Suite 200, Lanham, Maryland 20706
www.rowman.com

Unit A, Whitacre Mews, 26-34 Stannary Street, London SE11 4AB

British Library Cataloguing in Publication Information Available

Library of Congress Cataloging-in-Publication Data

Names: Rawlins, Ben, 1982– author.
Title: Mobile technologies in libraries : a LITA guide / Ben Rawlins.
Description: Lanham : Rowman & Littlefield, [2016] | Series: Library Information
Technology Association (LITA) guides | Includes bibliographical references and index.
Identifiers: LCCN 2016014799 (print) | LCCN 2016031672 (ebook) | ISBN
9781442264236 (hardback) | ISBN 9781442264243 (paperback) | ISBN 9781442264250
(eBook) | ISBN 9781442264250 (electronic)
Subjects: LCSH: Mobile communication systems—Library applications. | Libraries—
Information technology. | Library Web sites—Design.
Classification: LCC Z680.5 .R393 2016 (print) | LCC Z680.5 (ebook) | DDC 025.0422—
dc23
LC record available at https://lccn.loc.gov/2016014799

Printed in the United States of America

In memory of Mom—
you will always be with me.

Contents

List of Figures

Preface

Mobile technology has expanded to become an essential part of our lives in less than a decade. The introduction of the iPhone in 2007 opened the flood gates for the mobile revolution that would change the way we consume information and connect with the world around us. Shortly after the introduction of the iPhone, Apple and then Google released what would become the two most popular mobile operating systems in the world: iOS and Android. As devices like the iPhone and other smartphones began to gain more traction and adoption among consumers, businesses started to shift focus to the mobile platform to reach users in this ever-increasing medium. Apple has become a company whose success is built on mobile hardware and software. Google brings in a ton of revenue from mobile ads. Facebook has expanded its reach in mobile with mobile ads and the acquisition of mobile-only companies like Instagram.

In addition to the changes in business, mobile has changed the way that we develop websites. It has led to the rise of responsive web design, which is a mobile-first approach to designing websites. This has brought about the introduction of popular responsive frameworks such as Bootstrap. Wearable technology is starting to emerge as a serious category within the realm of mobile technology. All signs point to continued expansion of mobile technology.

With the expansion of mobile, libraries have seized the opportunities that mobile technology offers and that are the premise of this book. With more and more of our users coming into the library equipped with mobile devices, it is essential that libraries provide services and resources and develop the skills necessary to reach this user base. Libraries have done a good job of this, but with the continued expansion of mobile technology there are more opportunities to come. This book explores different ways that libraries have incorporated mobile technology in their services to reach their users. Some topics discussed in this book include

* the impact of mobile technology
* mobile technology and library services
* responsive web design
* wearable technology
* the future of mobile technology

Mobile Technologies in Libraries: A LITA Guide is written for library staff interested in how mobile technologies have changed the way we access, and expect to access, information, as well as how libraries can incorporate and adapt to mobile technology. This book consists of eight chapters:

* Chapter 1, "The Impact of Mobile Technology," looks at the overall impact of mobile technology and how companies like Apple, Google, and Facebook have seized the opportunities of the mobile revolution to help build their empires.
* Chapter 2, "The Digital Divide, Libraries, and Mobile Devices," discusses the digital divide and how libraries are working to bridge the divide. This chapter also looks at the role mobile technology plays in helping to decrease that divide.
* Chapter 3, "Mobile Technology and Library Services," discusses how libraries have incorporated mobile technology into services and how they are reaching mobile users. This includes a discussion on why mobile is important for libraries and examples of mobile websites, mobile applications, and other services, like lending devices, that libraries are providing to their users.
* Chapter 4, "Social Media, Mobile Technology, and Libraries," discusses the role of social media and how libraries are using various social media outlets to reach users.
* Chapter 5, "Responsive Web Design," focuses on how mobile technology has changed the way that we develop websites. It has led to the rise and popularity of responsive web design and the creation of responsive frameworks such as Bootstrap.
* Chapter 6, "What Library Vendors Are Offering," discusses what library vendors are providing in terms of mobile solutions. Some vendors are focusing on developing stand-alone mobile websites and responsive websites, others are developing mobile applications for both iOS and Android, and some vendors are developing both mobile web interfaces and mobile applications.

- Chapter 7, "Wearable Technology," talks about this emerging category within mobile technology. It focuses on the devices and in some cases how libraries are integrating wearable technology such as Google Glass, Oculus Rift, and GoPro cameras.
- Chapter 8, "The Future of Mobile Technology," looks at mobile technologies that are starting to garner attention, such as beacons and the Internet of Things. It explores what the future for mobile technology may consist of with some insight into how libraries might incorporate these future technologies.

Libraries are doing some great things in regard to mobile technology, and as it continues to expand, mobile technology offers librarians and libraries a host of opportunities to reach users and to show the value of our institutions and profession. I hope that you find this book to be a useful discussion on the ways that mobile technology has had a profound societal impact as well as a resource for ideas to better serve your users based on what other libraries are doing.

Acknowledgments

I would like to thank my colleagues at the Ensor Learning Resource Center (Michele Ruth, Andrew Adler, Randall Myers, Ernie Heavin, and Brandi Duggins), who have been a great support through this project. They are a great team who have engaged in many conversations with me regarding the value of mobile technology in libraries. They have worked on various projects with me to make our resources and services mobile friendly. Most of all I would like to thank my wonderful wife, Julia, whose support and encouragement helped make this project possible.

Chapter One

The Impact of Mobile Technology

The landscape of mobile technology changed forever when on January 9, 2007, Steve Jobs walked out onto the stage at the Macworld convention to announce the release of the iPhone. While it is true that smartphones existed before Apple's announcement, the iPhone was different because it gave us as consumers a glimpse into the future of mobile technology. While the first iPhone wasn't perfect, it was a device that allowed people to make phone calls, browse the web, check e-mails, and send text messages, among other things. It was a device that had its own operating system developed specifically with mobile in mind. Later that same year at Apple's Worldwide Developers Conference in June, Apple made another announcement: that the iPhone would support third-party apps from external developers. In 2008, Google followed suit with the release of its own open source mobile operating system, Android. It is hard to believe that in less than a decade, mobile technology has come to play such a vital role globally. In some ways it is even harder to imagine not having access to the various mobile devices on the market today as we have come to rely on them so greatly both personally and professionally. As a result, mobile technology has positioned itself as an essential aspect of our society that individuals and businesses are relying on more heavily each day.

The piece of mobile technology that has become integrated the most in our everyday lives is the smartphone. Smartphones have become part of who we are, and some of us cannot imagine being without them. Smartphones shape how we interact with the world around us and provide us with constant access to information no matter where we are (for the most part). The path to a device like the iPhone was a long time in the making. Humans are fascinated with technology, and we are constantly looking for ways to improve on the technology that is currently available. The road to smartphones can be traced back to 1973. On April 3, 1973, the first ever call was made using a cell phone. That call was made by Martin Cooper, a vice president and division manager at Motorola, to the head of research at Bell Labs during a press conference from Cooper's office in New York. Mazda paid homage to this event in one of its ads for the Mazda 3 in 2014. Mobile phones and their capabilities have changed dramatically since 1973, and global ownership and access to these devices have gone through the roof. Take, for example, a chilling statistic from the United Nations regarding mobile phone ownership and sanitation. A 2013 report from the United Nations stated that of the world's population of 7 billion people, 6 billion had mobile phones while only 4.5 billion had access to toilets and latrines.[1]

Along with smartphones, tablets have become an important piece of mobile technology; both devices will be discussed next.

SMARTPHONES

Smartphones continue to drive the current mobile landscape and have reached a point, at least in the United States, where a majority of the population owns them. According to Nielsen, more than two-thirds of Americans own a smartphone. Smartphone penetration is over 50% for nearly all age categories, with the exception of those over 65. Smartphone ownership according to age breaks down as follows:

- Ages 18–29: 85% smartphone ownership
- Ages 30–49: 79% smartphone ownership
- Ages 50–64: 54% smartphone ownership
- Age 65+: 27% smartphone ownership[2]

These devices have become extremely popular because they offer the user a variety of different capabilities and options, such as

- text
- calling
- e-mail
- entertainment
- navigation
- information
- job search
- photography
- news

- videography
- banking
- real estate
- investing
- social media
- dating[3]

Not only do smartphones offer users a host of different capabilities, but in some cases, they are the only means for Internet access. The Pew Research Center reported that 10% of Americans who own a smartphone do not have any other form of high-speed Internet access at home beyond their smartphone's data plan. Additionally, 15% of Americans who own a smartphone say that they have a limited number of ways to get online other than their smartphone.

The Pew Research Center found that certain groups of Americans rely more heavily on smartphones for online access:

- Younger adults—15% of Americans between the ages of 18 and 29 are heavily dependent on smartphones for online access.
- People with low household incomes and levels of educational attainment—13% of Americans with an annual household income of less than $30,000 are more reliant on smartphones for online access, while just 1% of Americans with an annual household income of more than $75,000 rely on their smartphones for online access to a similar degree.
- Minorities—12% of African Americans and 13% of Latinos are smartphone dependent for online access, compared to 4% of Caucasians.[4]

From these numbers, the Pew Research Center found that smartphone owners who are more reliant on their devices for online access are less likely to have another type of computing device.

Smartphone owners are relying more and more on their devices for important life events. Aside from using these devices for normal tasks such as calling, texting, and web browsing, smartphone owners are using their devices to navigate of variety of different life events:

- 62% of smartphone owners have used their phone in the past year to look up information about a health condition.
- 57% have used their phone to do online banking.
- 44% have used their phone to look up real estate listings or other information about a place to live.
- 43% have used their phone to look up information about a job.

- 40% have used their phone to look up government services or information.
- 30% have used their phone to take a class or get educational content.
- 18% have used their phone to submit a job application.[5]

Smartphones play an important role in helping users follow news and events that are occurring locally, nationally, and globally and in helping users share information with others:

- 68% of smartphone owners use their phone at least occasionally to follow along with breaking news events, with 33% saying that they do this "frequently."
- 67% use their phone to share pictures, videos, or commentary about events happening in their community, with 35% doing so frequently.
- 56% use their phone at least occasionally to learn about community events or activities, with 18% doing this "frequently."[6]

Users also use their smartphone to help them navigate and get them where they need to go. A majority of smartphones now come equipped with some type of GPS application, such as Apple Maps or Google Maps, that enables a smartphone to be used as a GPS for turn-by-turn directions:

- 67% of smartphone owners use their phone at least occasionally for turn-by-turn navigation while driving, with 31% saying that they do this "frequently."
- 25% use their phone at least occasionally to get public transit information, with 10% doing this "frequently."
- 11% use their phone at least occasionally to reserve a taxi or car service. Just 4% do so frequently, and 72% of smartphone owners never use their phone for this purpose.[7]

I can personally attest to the value of using smartphones as a GPS. I once used my iPhone as a GPS for a 7,000-mile cross-country trip, and the directions provided by Google Maps were just as good as a stand-alone GPS. Even makers of stand-alone navigation systems, such as TomTom, have developed navigation apps that can be used on smartphones.

The use of smartphones, and mobile devices, has become an important component in our lives. To see the importance of mobile devices, consider the results from a 2015 survey by the mobile technology advertising company the Mobile Majority. Responses from their survey included the following:

- 85% of survey respondents said that mobile devices are a central part of everyday life.
- 89% said that mobile devices allow them to stay up-to-date with loved ones and social events.
- 70% said that their smartphones represent "freedom."
- 46% said that their smartphone is something they "couldn't live without."
- 44% said they've had trouble doing something because their phone wasn't with them.[8]

Tablet computers are other mobile devices that have seen steady growth over the last few years.

TABLETS

Since 2010, tablet computers have seen substantial growth among users in the United States. Tablet computers, simply referred to as tablets, can be defined as general-purpose portable computers that are contained in a single touch-screen panel. Microsoft was the first company to use the term *tablet* when it built a device in 2001. Bill Gates, then CEO of Microsoft, was confident that tablets were the future of computing. He stated in an interview with CNN that "the tablet takes cutting-edge PC technology and makes it available when-ever you want it, which is why I'm already using a tablet as my everyday computer. . . . It's a PC that is virtually without limits—within five years I predict it will be the most popular form of PC sold in America." His predic-tion turned out to be a few years off. Microsoft went on to produce tablets, but other than in a few specialized sectors, the tablet market never took off in the mainstream market due to concerns from consumers. Among those concerns were the premium price for tablets over traditional notebooks, which made them cost prohibitive; tablets at the time ran on an operating system designed around the use of a mouse and keyboard; and they required the use of an ex-ternal input device such as a stylus.[9] As a result of these concerns, the market for tablets died down until the introduction of Apple's iPad in 2010.

Since the introduction of the iPad, tablet ownership has increased steadily. Steve Jobs said the introduction of the iPad was ushering in what he referred to as the "post-PC era." The iPad became the fastest selling electronic device that was not a phone when it sold more than 3 million units in its first 80 days.[10] According to the Pew Research Center, tablet ownership in the United States in 2010 amounted to 3% of the population. In 2015, tablet ownership jumped to 45%.[11]

Tablets have found their way into education. School districts, along with individual consumers, are seeing the benefits that these devices provide. Some school districts and higher education institutions around the country have implemented tablets and have looked for additional ways they can be incorporated into the curriculum. School districts like Brookfield High School in Connecticut, Burlington High School near Boston, and Woodford County High School in Kentucky have incorporated iPads into their curriculum. In 2011, one year after the release of the iPad, Apple reported that more than 600 school districts had launched a "one-to-one" program where at least one classroom was getting iPads for each student to use throughout the day.[12] Institutions in higher education have launched similar programs. In 2013, Lynn University in Florida and Arkansas State University launched initiatives that provided all incoming freshmen with an iPad or required students to purchase or rent one shortly after arriving on campus.

The future of the tablet is interesting. There is definitely room for growth, but there is not the same upgrade cycle that exists with smartphones. According to CNET, the future of the tablet is the PC. So what does that mean? It means that mobile computing in regard to tablets may be shifting to devices that serve as hybrids. Devices that can be used as both a tablet and a laptop are becoming more popular. Shipments for devices like Microsoft's Surface Pro 3 and Lenovo Yoga 3 Pro are expected to grow significantly in 2016.[13] In 2012, Apple CEO Tim Cook compared these hybrid devices to mashing up a refrigerator with a toaster, but Apple has since changed its tune slightly with the introduction of the iPad Pro and accessories that help it serve in some ways as a hybrid. Regardless of how this all plays out, people are looking more to mobile technology for their information and productivity needs. As a result of the increased proliferation and reliance on mobile technology, companies like Apple, Google, and Facebook have made mobile a number one priority.

APPLE, GOOGLE, AND FACEBOOK

Although smartphones have been around since 1997, the development and release of the iPhone in 2007 provided the launching pad for how essential mobile technology would become. Apple, Google, and Facebook are three companies that have played, and continue to play, a significant role in the mobile landscape.

Apple

While the first iPhone would not be viewed as anything special by today's standards, the introduction of the device opened the eyes of the world to what was possible with a mobile device (and also increased expectations). The iPhone showed people that many of the tasks that we complete on a desktop computer were possible in the mobile realm as well. The iPhone allowed users to browse the web, check e-mail, and receive text messages, among other things. It was not merely a phone with some added features; it was a mobile computer that fit in your pocket. It was built with an operating system, iOS, that was designed for use on mobile devices. Since the release of iOS, Apple has created a new programming language called Swift for developers to use to build native applications. This new programming language is designed to be used for iOS, OS X, watchOS, and tvOS. According to Apple, Swift builds on the best of C and Objective-C, without the constraints of C compatibility. It is meant to be a user-friendly programming language that is easy for new programmers to the Apple ecosystem to pick up.

The iPhone, and other devices that followed from Apple and other mobile device manufacturers, changed the way that we interact with the world around us. The iPhone enabled users to connect to and access information over a cellular network. At the time of the iPhone's release, Steve Jobs, cofounder and then CEO of Apple, said, "Every once in a while a revolutionary product comes along and changes everything."[14] While that quote may appear to be a marketing mechanism for Apple and its new product (and it was), it was an accurate statement. The iPhone did revolutionize things. It changed our expectations regarding mobile technology; it changed our information consumption behaviors; it changed the expectations we have of businesses and organizations; it changed the nature of web development; and it provided businesses and organizations with an additional medium for reaching people. Apple provided a successful blueprint for the smartphone, and later the tablet with the introduction of the iPad, that companies like Samsung would follow.

Following the release of the iPhone, Apple announced the App Store in 2008 and shortly thereafter opened up its iOS platform to third-party developers through the development and release of a native software development kit. However, third-party applications were not originally supported on the iPhone because Steve Jobs thought that developers could leverage Web 2.0 technologies to create web applications that behaved like native applications. Since the announcement of the App Store, users have downloaded more than 50 billion apps. Apple changed the tablet market in 2010 with the release of the iPad, which is currently Apple's second biggest product behind the iPhone.

Apple has benefited greatly from the mobile revolution, and a huge portion of its revenue is driven by mobile devices. According to figures from Apple (Figure 1.1), in the first quarter of 2015, sales of the iPhone and iPad made up 78% of its total revenue. Fast-forward just one year to the first quarter of 2016 and that number is nearly identical, with the iPhone and iPad constituting 77% of Apple's revenue. With the release of the Apple Watch it will

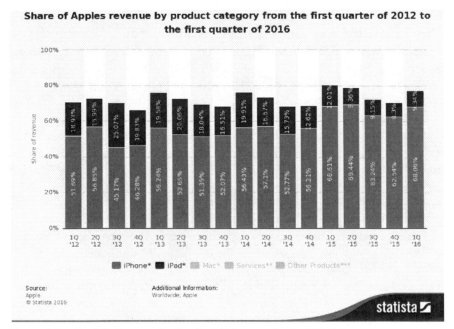

Figure 1.1.   Apple's Revenue by Product Category. *Apple / Statista.com*

be interesting to see how much of the market share Apple can obtain in the wearables category.

Google

Following Apple's lead and developing an operating system designed to run on mobile devices, Google released the Android operating system in 2007. The first Android-powered device was released in 2008. The Android operating system is a Linux-based system designed to run on mobile devices. It is primarily written in the Java programming language. A major difference between iOS and Android is that when Google released the code for Android, it did so as open source under the Apache License. Apple does not release the code to iOS to any developers outside of its company. The release of Android as open source allows the software to be freely distributed and modified by device manufacturers, wireless carriers, and developers outside of Google. While iOS only runs on Apple devices, Android runs on a variety of devices from manufacturers like Samsung, HTC, LG, and Sony. Amazon's Kindle Fire also runs on a customized version of Android. It is by far the most popular mobile platform worldwide: Android powers more than 80% of smartphones globally (Figure 1.2). Like Apple, Google has its own app

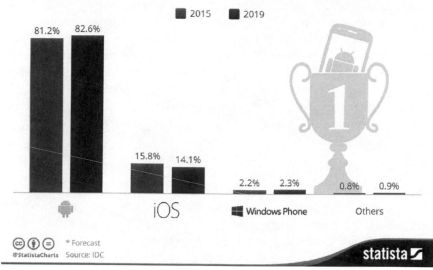

The Platform War Is Over and Android Won

Worldwide smartphone operating system market share (% of new device shipments)*

■ 2015 ■ 2019

81.2% 82.6%

15.8% 14.1%

2.2% 2.3% 0.8% 0.9%

iOS Windows Phone Others

* Forecast
@StatistaCharts Source: IDC

statista

Figure 1.2. Android Operating System Market Share. *IDC / Statista.com*

store, Google Play, where Android developers can upload their applications for sale.

Aside from Android, Google plays a significant role in the mobile technology in a variety of ways. Mobile advertising is one. Google generates a significant amount of revenue in this arena. In 2014, Google had $11.8 billion in mobile search ad revenue with 75% of that revenue being generated by iOS users.[15] Google is the leader in mobile advertising revenue. The company has a line of mobile devices called Nexus that are popular with Android purists, and it has developed a slew of mobile apps, such as Google Maps, Gmail, and Google Drive, that are available on both Android and iOS.

Another way that Google has made its mark in the mobile landscape is by changing its search algorithm to prioritize mobile-friendly websites on searches performed with a mobile device. According to an article in the *Wall Street Journal*, "Google said it tweaked its algorithm for mobile searches to favor sites that look good on smartphone screens, and penalize sites with content that is too wide for a phone screen and text and links that are too small." With mobile search becoming more popular among users, it is important that the websites Google relies on be mobile friendly. This move is important not only for end users but for Google as well. With mobile search starting to overtake search done on a desktop, Google also wants more users to surf the web on their phones instead of using mobile apps. Google sells ads that point to websites but generally cannot direct searches to content inside apps.[16]

Google has also acquired several mobile-related companies such as Waze. Waze is a community-based traffic and navigation app that was acquired by Google in June 2013 for $1 billion. Google has integrated some features of Waze, such as real-time user traffic reports, into its Google Maps application for both iOS and Android. The company has also invested some resources into the potential for mobile payments with Google Wallet. As mobile technology continues to grow and shape our interactions, expect companies like Google and Apple to have a major impact.

Facebook

There are other companies besides Apple and Google that have benefited from the mobile revolution. Facebook would be another example. Facebook's financial reports from the last quarter of 2015 show that it had 934 million mobile daily active users, with that number expected to eclipse 1 billion in the first part of 2016. Facebook has acquired some notable companies to help build on its mobile offerings (Figure 1.3). In April 2012, Facebook purchased Instagram, the popular photo sharing application, for $1 billion. In 2014, Facebook purchased WhatsApp, a mobile messaging application, for $19 billion.

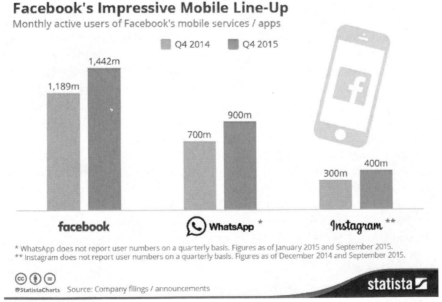

Figure 1.3.   Facebook's Mobile Lineup. *Company filings/announcements / Statista.com*

The company also began generating a significant amount of revenue via mobile advertising (second only to Google). Mobile advertising accounts for 78% of Facebook's total advertising revenue. In the third quarter of 2015, mobile advertising generated $3.4 billion of its $4.3 billion total advertising revenue.[17] Facebook had the foresight to recognize that there was huge potential in mobile advertising and has continued to grow its revenue from mobile advertising. There are reports that Facebook is considering adding ads to its mobile messaging app, Messenger. When and if that happens remains to be seen, but it is clear that Facebook has seized the opportunity that mobile has to offer.

MOBILE AND THE CONSUMER

With the success that both Google and Facebook have had in mobile advertising, companies are looking at the behavior patterns of mobile users to determine how they can actively engage consumers through their mobile devices. The Mobile Majority report "The 2015 Mobile Consumer" stated that the mobile consumer represents the holy grail of digital marketing and advertising for brands. The reason for that is that mobile is now the first screen for consumers even before desktops and television. Consumers now spend 2.8 hours, on average, a day interacting with their mobile phones. The report found that consumers use their devices from nearly any location:

- 99% of smartphone owners use their phones at home.
- 82% of smartphone owners use their phones while they are in a car or on public transit.
- 69% of smartphone owners use their phones while they are at work.
- 53% of smartphone owners use their phones while they are waiting in line.
- 51% of smartphone owners use their phones while at a community place.
- 50% of smartphone owners use their phones while walking from place to place.[18]

Understanding where and how consumers interact with their smartphones can help businesses develop strategies to reach and engage consumers on their mobile devices.

The Mobile Majority report also found that consumers show a willingness to interact and engage with a brand's content through their smartphones. The report found that

- 76% of consumers who use location sharing do so because it provides them with more meaningful content.

- 80% of consumers said that the reason they subscribe to e-mails from brands is for the coupons and deals.
- 65% of consumers who downloaded business-specific apps do so because they provide more convenient access to information.
- 77% of consumers who opt in to brand text messages do so because of the coupons and deals.
- 63% of consumers who follow brands on social media do so because they were offered coupons and deals if they followed or liked the brand's page.[19]

However, once brands get consumers involved in the mobile marketing campaign they need to ensure that they are not overwhelming or underwhelming consumers. The "2016 Mobile Consumer Study" by Vibes found the following reasons for consumers to unsubscribe from a brand or company:

- 59% said they received too many messages or updates.
- 51% said the information they received was not relevant.
- 41% said the coupons or incentives were not good enough.
- 34% said the messages they received were not timely.
- 20% said they could not personalize the information they received.
- 9% said they received too few messages.[20]

During the 2015 holiday season smartphones played a big role in the shopping behaviors of consumers. They used their smartphones in a variety of different ways, such as getting store locations, reading reviews, and comparing prices (Figure 1.4). With these insights, companies can develop and implement mobile strategies to reach consumers in the mobile environment. With consumers relying more on mobile technology, it is essential that companies make mobile a priority and understand how consumers are using mobile.

What does all of this mean for libraries? In short, a lot. With the explosion of mobile technology, libraries have an opportunity to reach users in a new and exciting format. Not only is it an opportunity, but it is a necessity that libraries and librarians maintain a keen awareness of the mobile behavior of our users so we can identify the types of mobile technology to make available to them and the types of services and resources that we should provide. Throughout the rest of this book we will explore how mobile has come to shape society in various ways and how libraries can take and are taking advantage of what mobile technology has to offer.

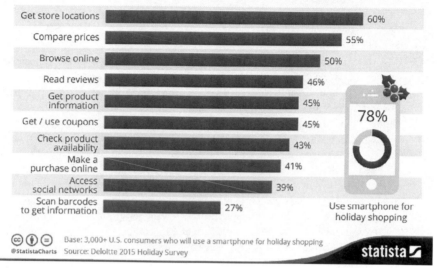

Figure 1.4. Smartphones and Holiday Shopping. *Deloitte 2015 Holiday Survey / Statista.com*

NOTES

1. United Nations. "Deputy UN chief calls for urgent action to tackle global sanitation crisis." United Nations, March 21, 2013. http://www.un.org/apps/news/story .asp?NewsID=44452#.VyIHK_krKUk

2. Mobile Majority. "The 2015 Mobile Consumer Report: Understanding Today's On-the-Go Buyer." Mobile Majority, 2016. http://www.majority.co/wp-content/uploads/2015/10/2015-Mobile-Consumer-The-Mobile-Majority-1.pdf

3. Ibid.

4. Smith, Aaron. "U.S. Smartphone Use in 2015." Pew Research Center, April 1, 2015. http://www.pewinternet.org/2015/04/01/us-smartphone-use-in-2015/

5. Ibid.

6. Ibid.

7. Ibid.

8. Mobile Majority, "The 2015 Mobile Consumer Report."

9. Griffey, Jason. "The Rise of the Tablet." *Library Technology Reports* 48, no. 3 (April 2012): 7–13.

10. Melloy, John. "iPad Adoption Rate Fastest Ever, Passing DVD Player." CNBC, October 4, 2010. http://www.cnbc.com/id/39501308

11. Anderson, Monica. "Technology Device Ownership: 2015." Pew Research Center, October 29, 2015. http://www.pewinternet.org/2015/10/29/technology-device-ownership-2015/

12. Associated Press. "Many U.S. Schools Adding iPads, Trimming Textbooks." *USA Today*, September 2, 2011. http://usatoday30.usatoday.com/news/education/story/2011–09–03/Many-US-schools-adding-iPads-trimming-textbooks/50251238/1

13. Statt, Nick. "The Future of the Tablet is the PC." CNET, August 22, 2015. http://www.cnet.com/news/the-future-of-the-tablet-is-the-pc/

14. Honan, Matthew. "Apple Unveils iPhone." *Macworld*, January 9, 2007. http://www.macworld.com/article/1054769/iphone.html

15. Oliver, Sam. "Apple's iOS drives 75% of Google's Mobile Advertising Revenue." *Apple Insider*, May 27, 2015. http://appleinsider.com/articles/15/05/27/apples-ios-drives-75-of-googles-mobile-advertising-revenue

16. Winkler, Rolfe. "Google Gives Boost to Mobile-Friendly Sites; Google Tweaks Search Algorithm to Favor Sites That Look Good on Smartphone Screens." *Wall Street Journal*, April 21, 2015.

17. Sullivan, Mark. "Facebook Now Makes 78 Percent of Its Ad Revenue on Mobile." *Venture Beat*, November 4, 2015. http://venturebeat.com/2015/11/04/facebook-now-makes-78-percent-of-its-ad-revenue-on-mobile/

18. Mobile Majority. "The 2015 Mobile Consumer Report."

19. Ibid.

20. Vibes. "2016 Mobile Consumer Study." Vibes, 2016. http://www.vibes.com/resources/2016-mobile-consumer-report/

Chapter Two

The Digital Divide, Libraries, and Mobile Devices

Given the importance of computers in society, it is imperative for people to have access to adequate high-speed Internet connectivity. As individuals we rely on Internet access to shop online at places like Amazon, check our bank accounts, pay our bills, search and apply for jobs, and connect via social media, among many other things. As professionals we rely on Internet connectivity to perform vital functions of our jobs and to communicate with other professionals. In libraries, Internet access is essential for our patrons to be able to access our vast resources that are increasingly available online. Internet access is needed to look to see if a book is available at the library and to access research databases from vendors like EBSCO and JSTOR. In some libraries, Internet access is vital for the operation of the integrated library system (ILS) as more vendors move to web-based ILSs. Although there have been tremendous strides made to increase the access to high-speed Internet, a divide still exists. This is what is often referred to as the digital divide. You may be wondering, What exactly is the digital divide? Why is it important? Why are libraries concerned with helping bridge the digital divide? What do mobile devices have to do with the digital divide? These are important questions, and ones that we need to concern ourselves with as librarians.

THE DIGITAL DIVIDE

Prior to the 1990s the term *digital divide* often described the division between those who had telephone access and those who did not. However, with the rise of the Internet, the meaning of the term shifted in the late 1990s due to concerns by scholars, policy makers, and advocacy groups about the gap in access to information and communication technologies like computers, network hardware and software, cellular phones, and televisions just to name a few. The digital divide is now used to describe the gap that exists between individuals who have readily available Internet access and those who do not, particularly access to broadband Internet. This divide is based on various demographic factors, including education, geography, age, socioeconomic status, and a host of other factors. The U.S. government realizes the importance of broadband Internet access. The 2013 White House report "Four Years of Broadband Growth" from the Office of Science and Technology Policy and the National Economic Council, stated that "broadband access is an essential part of our economy . . . to create jobs and grow wages at home, and to compete in the global information economy, the delivery of fast, affordable and reliable broadband service to all corners of the United States must be a national imperative."[1] The United States has made great strides in broadband access. The report goes on to state that in 2000 only 4.4% of American households had access to a broadband Internet connection. By 2010 that number had increased to 68%.[2] However, there is still a large number of households that do not have access to broadband Internet.

Despite the inroads that have been made in the last 15 years to increase the access to broadband Internet, challenges still remain. The White House report cites three significant challenges: adoption, speed, and pricing. In terms of adoption of broadband in the United States, the White House report states that 72% of Americans use broadband in their homes. Given the size and geography of the United States, the adoption of broadband is lower than in some other nations with a comparable gross domestic product per capita.[3]

Demographic factors have had a significant impact on the adoption of and access to broadband Internet service. For example, Figure 2.1 shows the difference in Internet access among the U.S. population based on educational attainment. The higher the educational attainment, the more likely a household in the United States is to have Internet access if it is available. Socioeconomic status is another important indicator of access to adequate high-speed Internet, as is ethnicity.

According to the White House report, speed is another significant challenge that remains. This challenge is based more on geography because many urban areas have access to high-speed Internet, while many rural areas do not. Globally, the United States stills lags behind many Asian and European countries in terms of overall Internet speed provided to consumers. Currently, the Internet speed in the United States ranks 14th (Figure 2.2). Although the adoption of broadband and access to it is lower in the United States compared to other countries, the rate of adoption is increasing. As a result, the United States needs to prepare for the increasing need for faster Internet connections for consumers, businesses, and other important institutions. In fact, the Federal Communications Commission (FCC) is intending to solicit public

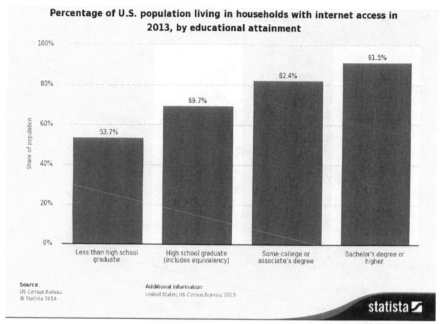

Figure 2.1. U.S. Internet Access by Educational Attainment. *U.S. Census Bureau /*
Statista.com

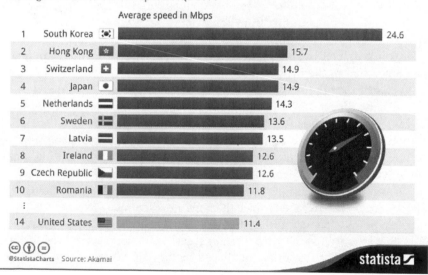

Figure 2.2. Countries with Fastest Internet Speeds. *Akamai / Statista.com*

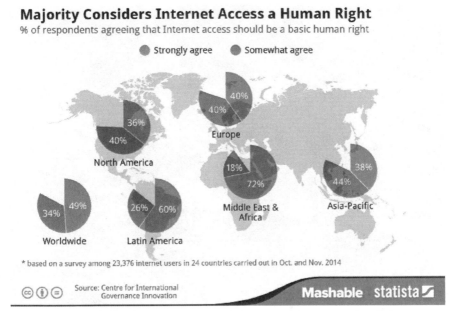

Majority Considers Internet Access a Human Right
% of respondents agreeing that Internet access should be a basic human right

● Strongly agree   ● Somewhat agree

40%
40%
36%
40%
Europe
North America
18%
72%
38%
44%
49%
34%
26% 60%
Middle East & Africa
Asia-Pacific
Worldwide   Latin America

* based on a survey among 23,376 internet users in 24 countries carried out in Oct. and Nov. 2014

Source: Centre for International Governance Innovation

Mashable   statista

Figure 2.3. Internet Access as a Human Right. *Center for International Governance Innovation / Statista.com*

feedback about the definition of *broadband*. Broadband Internet speed is currently defined at 4 megabits per second (Mbps). The FCC is determining whether that level should be more than doubled to 10 Mbps.[4] However, this has been met with opposition from companies like AT&T and Verizon, who argue that the current 4 Mbps connection is sufficient for users.

The price of Internet service, particularly broadband, is the third significant challenge cited in the White House report. According to the FCC in 2011, the average monthly price for a 1–5 Mbps connection was $35, the average monthly price for 5–15 Mbps was $44, and the average monthly price for a 15–25 Mbps connection was $56.50.[5] The digital divide has become more of a problem for lower income families now that access to high-speed Internet has become an essential tool for completing homework assignments at many public schools. In fact, federal regulators identified this gap in home Internet access as a key challenge in an education report in 2010. However, although Internet access has expanded since then, the Pew Internet and American Life Project reported that nearly a third of households with an income of less than $30,000 a year still do not have Internet access at home.[6] To further illustrate this point, consider the article published in the *Washington Post* about an eighth grader who would go to McDonald's to do his homework because it was one of the only places that offered free public Internet access after the local library closed.[7] With the increased importance of Internet access in our personal lives, as well as our educational lives, people are starting to view access to high-speed Internet as a basic human right (Figure 2.3). Additionally, almost half of Americans say that it would be very hard or impossible to give up the Internet (Figure 2.4). Americans would have an easier time giving up television and social media than they would the Internet, which is a very telling sign of how important the Internet has become in our society. These challenges are well documented, and there have been significant efforts to bridge the digital divide.

BRIDGING THE DIGITAL DIVIDE

Although the digital divide has been well documented and its challenges well known, it still exists to an alarming degree. Despite efforts to close the gap there is still much left that needs to be done. This is not just a case of a luxury that some people can afford and have access to and others cannot. Internet access has become so ingrained in the fiber of our society that it is not a luxury but a necessity. More and more, Internet access is needed for educational purposes, which is a burden on those students who do have access to broadband Internet at home. Speaking on the issue of broadband Internet access and

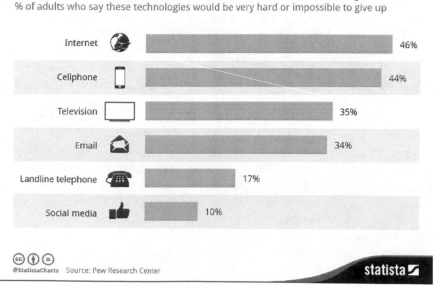

Figure 2.4.   Choosing Internet over TV. *Pew Research Center / Statista.com*

the digital divide, Susan Crawford, a professor at the Benjamin N. Cardozo School of Law and the former special assistant for science, technology, and innovation policy, stated in a 2013 interview that "parents around the country know that their kids can't get an adequate education without internet access."[8] She went on to state that "you can't apply for a job these days without going online. You can't get access to government benefits adequately. You can't start a business. This feels, to 300 million Americans, like a utility, like something that's just essential for life."[9] That is why it is important to continue to work on ways to bridge the digital divide. The benefits to closing the digital divide include improving democracy, social mobility, and economic growth, along with a host of others.

In the United States, democracy is at the foundation of our society. As citizens we have freedom of expression, freedom of the press, and freedom of assembly. These rights help us to keep our government accountable by affording us a voice. Our freedom of expression relies on having access to the free flow of information, quality education to adequately evaluate and analyze information, and open forums to engage in public discourse.[10] Increasingly the ability to engage in these types of activities relies on access to technology. Additionally, bridging the digital divide will enable politicians to interact more with the electorate through a variety of different channels and formats. Think of how successful the Obama campaign was in using the Internet, particularly social media, to reach out to voters. It could also mean increased participation in elections and a well-informed electorate.

Access to computers and high-speed Internet plays a vital role in the social mobility opportunities that exist in our society. With access to technology playing a more vital role in education, there is concern that students from lower socioeconomic statuses are placed at an unfair disadvantage, and thus are having their social mobility limited. Likewise, most jobs now require filling out and submitting an electronic application. Those without access to tools to accomplish this have their social mobility limited. As a result, there have been suggestions that the government look at ways to offer subsidies to help with this problem.

On a global scale, the digital divide has an effect on economic growth. In a January 2014 article, Rana Foroohar, a columnist for *Time* magazine, mentioned that the digital economy is growing at a faster rate than the real economy. With the lack of infrastructure and access to free speech, as well as red tape, in some developing countries, their economies are falling farther and farther behind because their citizens are unable to access the Internet easily, cheaply, and freely.[11] She also points out that policy makers in the United States must understand the need for better information technology infrastructure, and in terms of ease of Internet usage and commerce, the United States is falling behind. In order to foster economic growth in Lexington, Kentucky, Mayor Jim Gray is looking to improve on the sluggish Internet by seeking to build a fiber optic network and making Lexington a "gigabit" city. What this means is that the Internet speed would reach 1,000 Mbps. Currently, the average speed of the Internet connection in Lexington is 16.4 Mbps. The mayor's office has received many inquiries from potential businesses about Internet speed. Lexington's chief information officer, Aldona Valicenti, stated that "broadband availability will help and serve all of our citizens, businesses, students, entrepreneurs, and graduates from our colleges and universities . . . we have been assessing the situation; it's time to act."[12] This will be an expensive undertaking for the city and is estimated to cost around $200 million. It is something the mayor's office feels is necessary for the economic growth of Lexington and illustrates the point of how access, or lack of access, has a significant impact on economic growth and the potential for growth.

As a way to help bridge the digital divide and provide access to next-generation broadband, President Obama announced the ConnectED Initiative in June 2013. According to the White House, the ConnectED Initiative will connect 99% of American students to next-generation broadband Internet access in their schools and libraries by 2018. Following the announcement of the ConnectED Initiative and President Obama's 2014 State of the Union Address, it was announced that there had been $10 billion committed to the program including FCC funding for school and library connectivity and more than $2 billion in private sector funding from companies like Apple, Microsoft, Sprint, and Verizon, which helped to connect 20 million more students to high-speed internet access.[13] President Obama announced that the FCC would be making $8 billion available for the connectivity program. As of 2015, the FCC funding has brought high-speed connectivity to more than 10,000 schools and over 500 libraries across America.[14] Along with the FCC, the following private-sectors companies have also pledged support of more than $2 billion to provide cutting-edge technologies for classrooms:

- Adobe, which will provide more than $300 million worth of free software to teachers and students, including Photoshop and Premiere Elements for creative projects; Presenter and Captivate to amplify e-Learning; EchoSign for school workflow; and a range of teacher training resources
- Apple, which will donate $100 million in iPads, MacBooks, and other products, along with content and professional development tools to enrich learning in disadvantaged U.S. schools
- AT&T, which pledged more than $100 million to give 50,000 middle and high school students in Title I districts free Internet connectivity for educational devices over their wireless network for three years

- Autodesk, which pledged to make their 3D design program "Design the Future" available for free in every secondary school in the U.S.—more than $250 million in value
- Esri, which will provide $1 billion worth of free access to ArcGIS Online Organization accounts—the same Geographic Information Systems mapping technology used by government and business—to every K–12 school in America to allow students to map and analyze data
- Microsoft, which will launch a substantial affordability program open to all U.S. public schools by deeply discounting the price of its Windows operating system, which will decrease the price of Windows-based devices
- O'Reilly Media, which is partnering with Safari Books Online to make more than $100 million in educational content and tools available for free to every school in the United States
- Prezi, which will provide $100 million in Edu Pro licenses for high schools and all educators across America
- Sprint, which will offer free wireless service for up to 50,000 low-income high school students over the next four years, valued at $100 million
- Verizon, which announced a multiyear program to support ConnectED through up to $100 million in cash and in-kind commitments[15]

The White House and private-sector companies realize the importance of access to technology, in particular high-speed Internet access, and the impact that it has not only in the United States but on a global scale. Libraries have also been instrumental in the fight to bridge the digital divide.

LIBRARIES AND THE DIGITAL DIVIDE

Libraries have long been important educational centers. In addition to fulfilling this role, libraries, especially public libraries, are instrumental in helping to bridge the digital divide by providing access to high-speed Internet to patrons who would otherwise not have access. According to the Public Libraries and Internet Survey, conducted by the Information Policy and Access Center, public libraries are a key provider of high-speed Internet access. Key findings from the survey found that

- Almost all public libraries (99.3%) offer some form of free public Internet access
- Public libraries report being the only provider of free public access to computers and the Internet in 62.1% of communities in the United States
- 90.5% of public libraries offer free wireless access for users
- 99.1% of public libraries provide access to databases, and 81.9% provide access to homework resources

- 86.5% of public libraries report offering classes on general Internet use
- 87.0% of public libraries report offering classes in general computer skills
- 82.7% of public libraries report offering point-of-use technology training
- 44.3% of libraries offer formal technology training classes
- 34.8% offer one-on-one training sessions by appointment[16]

Libraries play a key role in offering high-speed Internet access to communities, and are instrumental in helping to build increased digital inclusion. Contrary to what some may believe, libraries are not becoming less relevant. They are becoming more important now than ever, as access to high-speed Internet and information continues to be an important and necessary aspect of our lives. Currently, public libraries are still the primary source of broadband access for many Americans. In two-thirds of American communities, public libraries are the only source of freely available Internet access that includes access to public computer terminals. In many rural areas, public libraries constitute the only means that some communities have to connect to high-speed Internet.[17]

Some libraries are looking to expand access to broadband Internet through various different programs, such as checking out broadband Internet hotspots to patrons. The New York Public Library system, for example, is looking to allow local residents with a library card to take home broadband Internet hotspots for up to a year. These devices allow users to connect up to 10 devices—such as laptops, tablets, and smartphones—to 4G LTE broadband. The hotspot devices normally cost about $49, powered by Sprint, with a two-year service agreement and a $110 monthly fee for 30 GB of data. New York libraries will lend hundreds of these devices to patrons with free unlimited data as long as they are enrolled in some type of library program. The program is called "Check Out the Internet" and was started because it was found that nearly 30% of households in New York City did not subscribe to broadband Internet at home.[18] Linda Johnson, president and chief executive officer of the Brooklyn Public Library, stated, "Too many Brooklyn residents are on the wrong side of the digital divide. Free Wi-Fi at local BPL branches is a vital resource, but it can't make up for the lack of internet access in the home—access that helps children succeed in school, and provides parents with critical information on health, employment, education, and more."[19] New York Public Libraries are also reaching out to state libraries in Kansas and Maine to see how a program of this nature could work in less populated areas.

Despite the inroads that libraries are making, and attempting to make, there are still challenges. There is a disparity between home broadband access in rural areas and home broadband access in urban areas. This creates a challenge for libraries in rural areas because more people come to the library looking for Internet access. The difficulty and cost of creating the infrastructure necessary in rural areas to provide and support broadband Internet access is a major concern. The profit motive for companies that provide broadband Internet access is minimal given that the costs associated with providing the infrastructure would not enable them to recoup their cost quickly.[20] This is not the only challenge that libraries face. Other challenges include lack of adequate staffing to provide patrons with the assistance needed to interact with technology, such as technology training and employment assistance. Another challenge is replacing public computer terminals. Many rural libraries replace public computer terminals when the need arises, while a majority of libraries in urban areas have a technology replacement plan. This comes down to an issue of funding that is a challenge not only in public libraries, but in all types of libraries. Typically, larger libraries have bigger technology budgets that enable them to have more up-to-date technology than smaller libraries. However, despite the challenges, libraries are playing a vital role in helping to bridge the digital divide.

MOBILE TECHNOLOGY AND THE DIGITAL DIVIDE

Mobile technologies are providing additional avenues by which the gap in the digital divide can be narrowed. Access to and adoption of mobile devices, particularly Internet-capable mobile devices such as tablets and smartphones, are continuing to increase. In some instances, these devices are the sole means that people use to connect to the Internet. Due to the continued rise in adoption and importance of mobile devices, IBM had some predictions back in 2011 about mobile technology in relation to the digital divide. In their annual "5 in 5" list, which they try to predict emerging technologies that will have a significant impact on our lives over the next five years, they predicted that the digital divide would cease to exist within five years due to advances in mobile technology. Paul Bloom, IBM's chief technology officer for Telco Research in 2011, stated, "In our global society, the wealth of economies are decided by the level of access to information. And in five years, the gap between information haves and have-nots will cease to exist due to the advent of mobile technology."[21] A bold prediction indeed, and one that, unfortunately, did not come true as the gap still exists. While mobile technology has increased rapidly and mobile devices have enabled better access to high-speed Internet, some significant challenges still exist (more on that in a moment). However, there have been tremendous strides in terms of mobile technology that have enabled more people to access high-speed Internet.

As more and more people continue to purchase and access the Internet using mobile devices, strain is put on mobile networks. Initially wireless companies struggled to keep up with the continued need and demand for access to the mobile networks and became overloaded with the requests that users were sending to them. As a result, connections at times could be extremely sluggish, and less frequently still are. There have been vast improvements to wireless networks as additional capital has been poured into enhancing the infrastructure to support wireless communication. This enhanced wireless infrastructure has led to 4G LTE networks that offer faster speeds. In some cases, enhanced wireless networks have been able to reach some rural areas and provide access to high-speed Internet through the use of mobile devices, areas where broadband service providers have not been able or willing to go.

Mobile devices, and technology in general, are playing a more important role in the educational system. As teachers and administrators realize the need for access to technology in order to acquire an adequate education, they are looking to develop and implement programs that will help bridge, or at least narrow, the digital divide among their students. One way they are doing this is through the use and deployment of mobile devices. For example, some school districts in the United States are rolling out iPad programs with the aim and getting an iPad in the hands of each of their students. McAllen School District in Texas and Woodford County School in Kentucky are two examples of schools that have implemented iPad programs. These school districts are hoping these iPad programs enhance student learning and engagement, and create more educational equity among their students. Some other districts are equipping their teachers with mobile devices to help educators understand the learning power and potential of these devices.

Even though mobile devices are instruments that have helped narrow the digital divide in the United States, they have not completely eliminated it, something that may never truly be done. With regard to the adoption of mobile devices, there is still some divide along income, age, and educational lines, much like it is with adoption of broadband Internet access. Additionally, mobile device users interact with the Internet differently than they would on a laptop or desktop computer. There are still some tasks and services that are easier to engage on a laptop or desktop.

Although mobile technologies present their own unique set of challenges, they are becoming essential aspects of our lives. They have altered the way that we think in many different areas and have had a tremendous impact on how we access information. The impact of mobile technologies, particularly mobile devices, cannot be overstated. One area of impact has been on library services.

NOTES

1. White House. *Four Years of Broadband Growth.* June 2013. http://www.white-house.gov/sites/default/files/broadband_report_final.pdf
2. Ibid.
3. Ibid.

4. Fung, Brian. "The FCC May Consider a Stricter Definition of Broadband in the Netflix Age." *Washington Post*, May 30, 2014. http://www.washingtonpost. com/blogs/the-switch/wp/2014/05/30/the-fcc-may-consider-a-stricter-definition-of-broadband-in-the-netflix-age/

5. Federal Communications Commission. *International Broadband Data Report*. August 2012. http://www.fcc.gov/document/international-broadband-data-report

6. Troianovski, Anton. "The Web-Deprived Study at McDonald's." *Wall Street Journal*, January 28, 2014. http://www.wsj.com/articles/SB10001424127887324731 304578189794161056954

7. Ibid.

8. Moyers & Company. *Moyers & Company: Who's Widening America's Digital Divide?* Films On Demand. Films Media Group, 2013.

9. Ibid.

10. Lamb, Jason. "The Digital Divide: Free Expression, Technology and a Fair Democracy." White Paper, Temple University, 2012. http://sites.temple.edu/lawdc-summer/files/2012/07/Digital_Divide_White_Paper_Jason_Lamb-use2.pdf

11. Foroohar, Rana. "The Real Threat to Economic Growth is the Digital Divide." *Time*, January 22, 2014. http://business.time.com/2014/01/22/the-real-threat-to-economic-growth-is-the-digital-divide/

12. Musgrave, Beth. "Mayor Gray Wants Lexington to Become a 'Giga-bit' City." *Lexington Herald Leader*, September 9, 2014. http://www.kentucky. com/2014/09/09/3420469/lexington-will-explore-partnership.html

13. White House. "ConnectED Initiative." June 2013. http://www.whitehouse. gov/issues/education/k-12/connected

14. White House. "FACT SHEET: ConnectED: Two Years of Delivering Opportunity to K-12 Schools & Libraries." June 25, 2015. https://www.whitehouse.gov/ the-press-office/2015/06/25/fact-sheet-connected-two-years-delivering-opportunity-k-12-schools

15. Ibid.

16. Information Policy and Access Center. "2011–2012 Public Library Funding and Technology Access Survey: Executive Summary." June 19, 2012. http://ipac. umd.edu/sites/default/files/publications/2012_plftasexecutivesummary.pdf

17. Real, Brian, John Carlo Bertot, and Paul T. Jaeger. "Rural Public Libraries and Digital Inclusion: Issues and Challenges." *Information Technology and Libraries* 33, no. 1 (March 2014): 6–24.

18. Scola, Nancy. "New York City Libraries Soon Will Let Patrons 'Check Out the Internet.'" *Washington Post,* December 4, 2014. http://www.washingtonpost. com/blogs/the-switch/wp/2014/12/04/new-york-city-libraries-soon-will-let-patrons-check-out-the-internet/

19. Ibid.

20. Real, Bertot, and Jaeger. "Rural Public Libraries and Digital Inclusion," 6–24.

21. Bloom, Paul. "IBM 5 in 5: Mobile Is Closing the Digital Divide." *IBM Research Blog*, December 19, 2011. http://ibmresearchnews.blogspot.com/2011/12/ ibm-5-in-5-mobile-is-closing-digital.html

Chapter Three

Mobile Technology and Library Services

The continued expansion, growth, and influence of mobile technology have given libraries the challenge, and opportunity, to develop, evaluate, and implement access to information in a format optimized for mobile devices. It is a challenge born out of both necessity and expectation. In order to remain relevant, libraries must continue to adapt to the changing information behaviors of their users. Additionally, users expect to have access to resources in formats that they desire. If they are unable to obtain that access, they will move on to other information access points that fit their needs and expectations. This challenge is not unique to libraries, and it is not an easy task. With the various types of mobile technology available and the costs associated with the technology, it can be difficult for a library to determine what the best route is to meet the needs and expectations of users. Despite the challenges, libraries have found ways to implement mobile services and resources to meet user expectations. Additionally, mobile devices have enabled libraries and librarians to rethink the way that traditional services, such as reference, can be offered.

Mobile technology provides tremendous opportunities for libraries. For example, mobile devices afford libraries an additional format to reach users and enhance the access to the vast amount of resources available. However, an additional challenge for libraries can be determining where to start. How do libraries know what types of mobile technology users are expecting and interacting with? In what ways are potential library users engaging with their mobile devices? Where is this information? Fortunately, there are a few avenues through which libraries can obtain this information in a broad sense. The Pew Research Center and Horizon Reports are both resources that track and focus on the use of mobile technology in different facets of society. Information from these resources can be very useful in determining the overall trends in mobile technology and how they might be applied in a particular library setting. Another option is to try to obtain information about the mobile usage behavior of your local patrons in order to tailor your mobile services for your particular audience. Many libraries are looking past the challenges that mobile technology presents and are making mobile a priority. Some ways that libraries are doing this include providing mobile interfaces, lending mobile devices, engaging social media, and reevaluating traditional services. To offer mobile services effectively, it is important that librarians understand the importance of mobile.

WHY MOBILE IS IMPORTANT FOR LIBRARIES

The growth and expansion of mobile technology has shifted how we access information. Given this shift in information behavior, mobile technology is an important consideration for libraries. The Pew Research Center and Library Edition of the Horizon Report illustrate the increasing importance of mobile in global society and for libraries, specifically academic and research libraries. The Pew Research Center is a nonpartisan organization that informs the general public about the issues, attitudes, and trends that are shaping global society. They conduct public-opinion polling, demographic research, media content analysis, and other empirical social science research, some of which has been focused on mobile technology usage and behavior. The Pew Research Center tracks and updates statistics on the percentage of American adults who own cell phones, smartphones, tablets, e-readers, and other behavior related to the use of mobile technology. According to their January 2014 "Mobile Technology Fact Sheet," statistics show that

- 90% of American adults own a cell phone.
- 58% of American adults own a smartphone.
- 42% of American adults own a tablet.
- 32% of American adults own an e-reader.[1]

Comparatively, in May 2011, statistics from the Pew Research Center show that ownership of these devices were as follows:

- 83% of American adults own a cell phone.
- 35% of American adults own a smartphone.
- 8% of American adults own a tablet.
- 12% of American adults own an e-reader.[2]

While overall cell phone ownership did not increase significantly, ownership of smartphones, tablets, and e-readers did. This information is useful for libraries and librarians, because we can see the impact and ownership of mobile technology is growing in such a short amount of time. Technology as a whole has become an important aspect of libraries, and as a result the New Media Consortium (NMC) released a Library Edition of the Horizon Report in 2014, devoted to examining the key trends, significant challenges, and emerging technologies that they expect to have a significant impact on academic and research libraries. Various aspects of mobile technology are an important aspect of this report.

The "NMC Horizon Report: 2014 Library Edition" was a collaboration between the New Media Consortium, University of Applied Sciences Chur, Technische Informationsbibliothek Hannover, and ETH-Bibliothek Zurich. One aspect of the report that deals with mobile technology and libraries is in the section titled "Trends Accelerating Technology Adoption in Academic and Research Libraries." In that section, they identified the prioritization of mobile content and delivery as a fast trend, which will have an impact on libraries in one to two years. They also identified mobile apps as important developments in technology for academic and research libraries whose adoption will be within one year or less.[3]

The Horizon Report states that the rise of mobile technology is changing scholarly workflows. Students and researchers have a stronger preference to search the library's catalog, read abstracts and full digital content, and find and save references through a library's mobile site or app rather than visit the physical library facilities.[4] As a result, libraries are looking at ways to prioritize content and delivery of resources into mobile-optimized formats that enable users to access this information from their mobile devices, thus increasing the productivity of students and researchers via their mobile devices. The report goes on to state that facilitating the movement toward mobile content and delivery requires leadership among library associations, professional development providers, and other academic and research libraries. The report also identifies mobile apps as a technology that will have a significant impact on libraries in 2015 and beyond.

Mobile apps have caused a dramatic shift in software development and have been popular forms of development. The Horizon Report indicates that mobile apps are an important technology that will continue to impact libraries. According to the report, mobile apps continue to gain traction in academic and research libraries because they are useful instruments for learning as they enable people to experience new concepts wherever they are, often across multiple devices.[5] More and more libraries are realizing the usefulness of mobile apps and are looking to either develop their own or outsource the development of a library app to a third-party vendor such as Boopsie. Information from both the Pew Research Center and the Library Edition of the Horizon Report illustrate the importance of mobile technology and the impact it has on libraries. Armed with this information and understanding, libraries are looking at ways to offer mobile services to reach mobile users. One way that libraries are accomplishing this goal is through the development and implementation of mobile-optimized interfaces.

MOBILE-OPTIMIZED INTERFACES

The continued increase of mobile web traffic has made it essential for libraries to provide some type of mobile-optimized interface for users to access resources. Being able to consume information in a mobile-optimized format is an expectation that many users have. Libraries realize this and are developing and implementing mobile-optimized interfaces such as mobile catalogs, mobile websites, and mobile applications to meet these expectations.

Mobile Catalog Interfaces

One way that libraries are providing mobile-optimized access is through the implementation of mobile library catalogs. Since users are expecting to be able to search the library's resources using their mobile devices, this provides a good option for libraries. Many integrated library system vendors offer mobile websites or mobile catalog skins that are available to subscribing libraries. For library systems that do not have this option, librarians develop a mobile skin for their catalog and share it with other member libraries via open source. An example of this would be the University of Rochester Libraries mobile Voyager skin created by Denise Dunham using the jQuery Mobile framework. Examples of libraries that have a mobile-optimized catalog interface include the University of Kentucky Libraries, California State University–Monterey Bay Library, and Jefferson County Public Library. At Georgetown College, we direct our mobile users to the mobile version of WorldCat Local.

Mobile Websites

Mobile websites offer users more than just being able to search the library's catalog. Users expect to be able to perform a wide range of tasks on their mobile devices and have the ability to access various forms of information. Libraries realize this, and rather than offering just the bare minimum—such as contact information, location, and library hours—mobile library websites are offering access to more information and resources. A survey conducted by MIT Libraries in 2011 found that people used or wanted to use their mobile devices to access library resources in the following ways:

- read e-books
- take notes
- read academic papers
- annotate academic papers or e-books
- listen to or watch lectures or podcasts
- search for library-owned books or journal articles
- renew library books
- request library books or articles[6]

Although this survey was conducted in 2011, it includes the same activities that faculty, staff, and students at Georgetown College, and many other libraries, today want to be able to do when they access library resources using their mobile devices.

Another study evaluated academic library mobile websites to see what items are most frequently used. The study found the following to be the most used features on academic library mobile websites:

- search link (86%)
- hours (80%)
- locations/maps (59%)
- contact us/contact information (53%)
- "Ask a Librarian" (47%)
- news/events (32%)
- personal account/renew (28%)
- search box (25%)
- research guide (by subject) (24%)
- laptop/computer availability (18%)
- study room reservation (11%)
- feedback (11%)
- social network (7%)
- FAQ/help (7%)
- staff directory (7%)
- about us (5%)
- course reserves (4%)[7]

There are several options available to libraries looking to develop a mobile website. Aside from building it from scratch, libraries can use a variety of popular mobile frameworks. One popular mobile framework that many libraries are using is the jQuery Mobile framework. This framework is a touch-optimized HTML5 framework that works across a majority of mobile platforms. It has been used for a variety of different projects in libraries—such as the

North Carolina State University Libraries WolfWalk project and Boise Public Library's mobile websites—and for the mobile version of WorldCat. However, as responsive web design (more on this in Chapter 5) becomes more popular, libraries are not developing stand-alone mobile websites. More and more libraries are transitioning mobile websites and the full library website to responsive layouts. Libraries are also targeting specific mobile platforms to reach mobile users through the development of mobile applications.

Mobile Applications

The popularity of the iOS and Android mobile platforms have led libraries to develop mobile applications to target these platforms. Developing native library applications can seem like a daunting task because each platform has its own programming language. In the case of iOS development, there are now two major programming languages: Objective-C and Swift. Both are object-oriented programming languages, but developers have noted that Objective-C requires a steep learning curve. In an effort to aid developers, Apple developed its own programming language, Swift, and announced it at the June 2014 Worldwide Developers Conference. Although Swift shares some similarities with Objective-C, it is supposed to be an easier language to pick up, one that is more concise and more resilient to errors in code. Swift can only be used to develop in the Apple ecosystem. Java, another object-oriented language, is the primary programming language for Android development. While there are other mobile platforms—such as Windows Mobile and Amazon Kindle—iOS and Android hold a significant portion of the mobile platform market share, and many libraries are focusing on targeting those platforms.

There are some distinct advantages to developing a native application for your library. For example, native applications

- have the ability to tap into device functionality such as the camera, GPS, push notifications, and other device features that are not currently possible to incorporate in a website
- can be used offline (although some Internet or data connection will be necessary if your application links out to resources not contained in the application)
- often provide a better user experience
- are more discoverable since users can search for them in Apple's App Store or on Google's Play Store

With users spending more time interacting with native applications, it makes sense that libraries are developing or outsourcing app development.

A search of the App Store and the Play Store show that many libraries have developed applications for both the iOS and Android platforms; examples include UCLA Libraries, Orange County Library System, New York Public Library, and Georgetown College. These applications offer users some of the same features they would find on their library's website. For example, features included on the New York Public Library's (NYPL) application include the ability to

- quickly search the collection, with filters to zero-in on the kinds of titles you're seeking
- browse best sellers and new material
- get the details on any title, anytime and anywhere, including descriptions, community reviews, and commentary
- check a title's availability—even map the NYPL locations where your title is available now
- place and manage holds
- renew items
- check location hours

At Georgetown College, we have implemented the following features into our mobile application (Figure 3.1):

- The ability to check the academic year library hours
- The ability to search for books, articles, journals, eBooks, and other resources in the library's catalog
- A listing of mobile research options such as the library's catalog, LibGuides, mobile-optimized databases, and mobile apps
- The ability to connect with a librarian to ask questions via chat, e-mail, and phone
- A directory of library staff that includes contact information and subject liaisons
- A "My Account" login to renew books or see when they are due
- A map of Georgetown College's campus
- Links to the library's social media accounts
- A link to the full library website

In addition to applications that offer users access to various library resources, some libraries have developed mobile applications focused on special projects and games. Two examples are North Carolina State University Libraries' WolfWalk application and Grand Valley State University Libraries Library Quest App. WolfWalk is a free location-aware iOS application that is

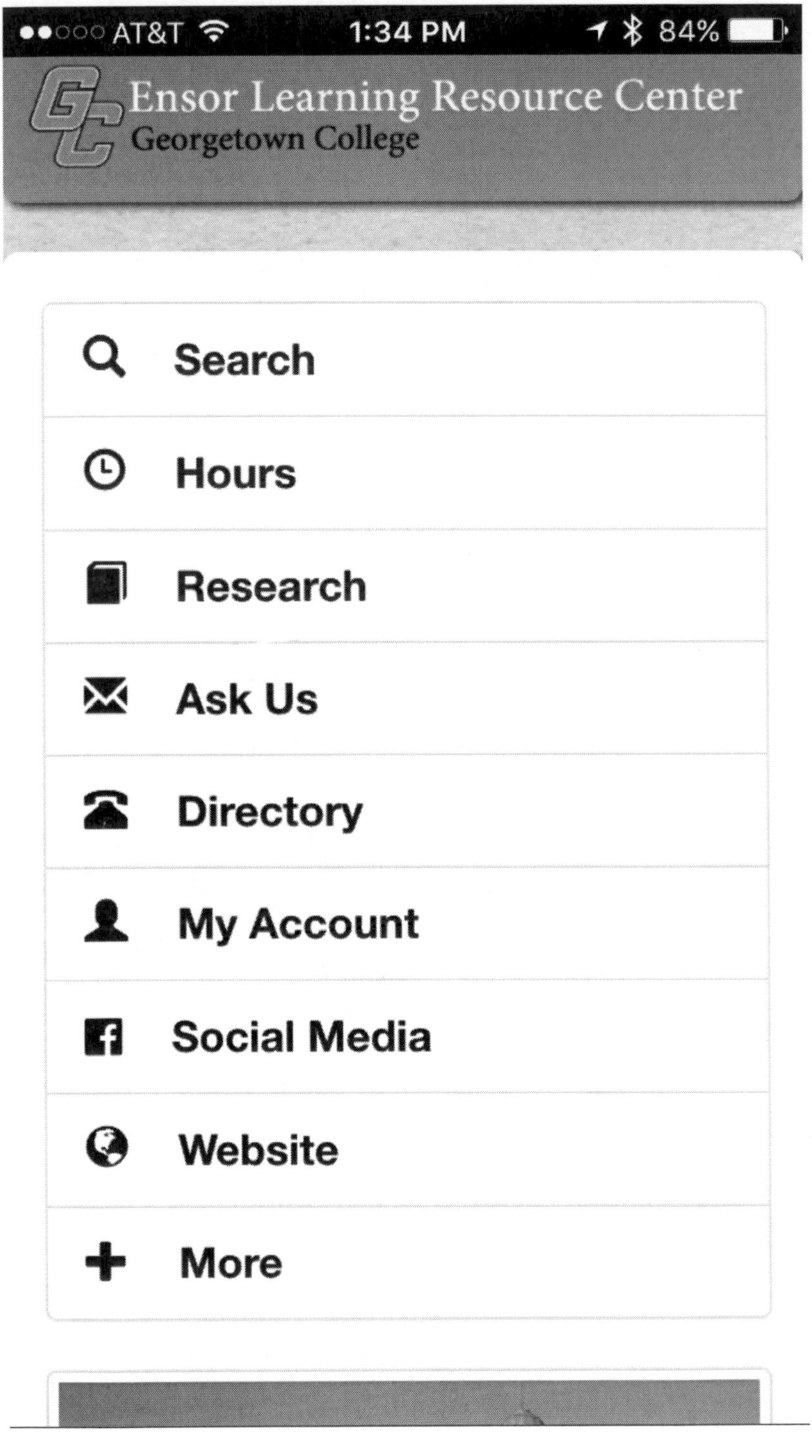

 Ensor Learning Resource Center
Georgetown College

🔍 **Search**

🕑 **Hours**

📕 **Research**

✉ **Ask Us**

☎ **Directory**

👤 **My Account**

📘 **Social Media**

🌐 **Website**

➕ **More**

Figure 3.1.   LRC Application

a photographic history of North Carolina State University that enables users to take a historical walking tour of campus and browse photographs by place, decade, or theme. Library Quest is a gaming application for iOS and Android, developed by YETi CGI, where library users complete library quests to earn points for chances to win prizes and earn library perks. The Horizon Report mentions that the best applications are tightly integrated with the capabilities of the device itself, using location data, motion detection, gestures, access to social networks, and web search, to seamlessly create a full-featured experience.[8]

Despite the advantages of developing a mobile library application, there are some drawbacks. One drawback is that native applications need to be developed for each specific platform. Since programming languages vary across each platform, that means the maintenance of multiple code bases. Mobile applications are also more time consuming and expensive to develop. Additionally, some libraries do not have the expertise and knowledge on staff to allow them to develop a mobile application. However, there are several frameworks that enable libraries, as well as other developers, to develop applications for various mobile platforms without having to learn that platform-specific programming language. Examples include IONIC, Mobile Angular UI, Sencha Touch, and PhoneGap.

IONIC, Mobile Angular UI, and Sencha Touch are HTML5 frameworks that allow developers to create applications for various mobile platforms. IONIC is considered to be one of the more promising frameworks. IONIC is built with Sass (Syntactically Awesome Style Sheets) and uses the JavaScript MVVM (Model-View-View Model) framework AngularJS to help develop interactive applications. Mobile Angular UI uses AngularJS in combination with Bootstrap to create mobile applications. Its main features include mobile components from Bootstrap such as switches, overlays, and sidebars. It also includes AngularJS modules such as angular-route, angular-touch, and angular-animate. The Sencha Touch framework comes with 50 built-in components and themes. It has a built-in Model-View-Controller and includes features like animations, DOM manipulations, AJAX, and touch events. A fourth framework is PhoneGap. Built on the Apache Cordova framework, PhoneGap is a mobile development framework that enables developers to use exiting web technologies, such as HTML, CSS, and JavaScript, to build application with native functionality through the use of application programming interfaces. It can be used with other frameworks such as jQuery Mobile and AngularJS to create mobile application for various mobile platforms. The use of these tools and frameworks has made the development of mobile applications easier for libraries and other developers. As mobile applications continue to be a major aspect of mobile usage among users, it is important for libraries to at least explore the option of developing a mobile library application. Regardless of what route your library takes, it is important to offer users some type of mobile access since they are expecting to have information available to them in the format of their choice.

In addition to developing and implementing mobile interfaces to reach mobile users, more libraries are lending mobile technology to users.

LENDING MOBILE DEVICES

While developing mobile solutions for users is a great way to extend the reach of your services and resources, it targets users who have mobile devices, who are only a segment of the population that any library serves. Not all patrons have access to a mobile device, have the ability to acquire a device, or are sure of the benefits of certain types of mobile technology. As a result, more and more libraries are starting to lend mobile technology to their users. By lending mobile technology, libraries are giving their entire user base the opportunity to interact with the technology and access the mobile solutions the library has created or has access to, in ways that users would have been unable to otherwise.

Although many libraries are now lending or plan to lend mobile devices, there is no consensus on the types of devices that are the best to circulate or strategies and policies related to lending. These are more localized decisions, where one library might experience a greater benefit from lending e-readers, while another might experience a benefit from lending tablets. Regardless of the approach, there are advantages and drawbacks to lending mobile technology to users.

There are a wide variety of examples of libraries lending mobile devices to users. To garner feedback from libraries that are lending mobile devices, three questions were posed to various library listservs.

1. Why did your library decide to lend mobile devices?
2. What devices are you lending?
3. What do you think are the benefits and drawbacks of lending mobile devices?

Several libraries responded and shared their insight. Among those that responded, and are lending mobile devices, were library staff from Marshall University Libraries, Temple University Libraries, Cooper-Siegel Community Library, Louisville Presbyterian Theological Seminary, and West Vancouver Memorial Library. Responding on their library's decision to lend mobile devices, they stated the following:

- "We lend these devices because people use these devices."
- "Students need to keep up with the technology their peers are using and to stay relevant in the job market."
- "We have a first generation student population who needs these technologies, but cannot necessarily afford to purchase them."
- "To support students doing work for courses or continuing after course enrollment to work on projects, including student films, website, and entrepreneurial activities."
- "To allow students and faculty to try out technology that they might be interested in purchasing personally, such as Kindles and iPads."
- "We saw that other libraries in our area were starting to lend mobile devices, and we did not want to be left behind."
- "The library thought it would be a good way for our patrons to try out different devices without having to buy them."
- "We decided to add the devices to our equipment lending because of student and staff interest."
- "Initially, once a device came to Canada, it was to help the community learn more about e-readers, but that has changed."
- "Kindles: bundles of curated content—to help highlight staff Reader's Advisory skills, and provide bundles of content to voracious readers. We have 8 adult genre Kindles with 15 titles each, a set of Kindles for Kids, and a set for Teens."
- "Kobo Load and Go—meeting a need to get e-books (especially born-digital content) into the hands of readers who do not have their own devices. We load up items from our Overdrive collection only."
- "iPads—this is part of a hands-on training program. The idea is that people need to practice skills to better learn them. Take the class and borrow the device."[9]

As you can see, there are a variety of reasons that libraries are lending mobile devices to their users.

These libraries are lending a variety of different mobile devices to their users. Marshall University Libraries lends iPads (first, second, and third generation). Temple University Libraries lends about two dozen iPads and four Kindles. However, the senior associate university librarian at Temple University, Jonathan LeBreton, added that they are no longer buying Kindles because they are not accessible to users with vision disabilities. Cooper-Siegal Community Library in Pittsburgh, Pennsylvania, lends Nook Colors, Nook Glows, iPad Minis, Kindle Fires, Kindle Paperwhites, and Google Nexus tablets. Louisville Presbyterian Theological Seminary is currently lending four Google Nexus 7 tablets. West Vancouver Memorial Library lends Kindles with preloaded content, Kobos that have OverDrive added e-books, and iPads. There are a host of other libraries that are lending mobile devices to users. For example, North Carolina State University Libraries; L.E. Phillips Memorial Public Library in Eau Claire, Wisconsin; University of Louisville Libraries; Oregon State University Libraries; and the Ensor Learning Resource Center at Georgetown College all lend mobile devices to users. At the Ensor Learning Resource Center, we currently lend six Kindles, which are loaded with popular titles such as the Game of Thrones and Twilight series, and six Nooks for users to download and access academic e-book titles. Our library has been considering lending additional mobile technology, but have yet to make the plunge into other devices such as tablets. There is not a shortage of libraries that have taken this plunge and have seen some significant benefits, and drawbacks, to offering these devices to users.

Lending mobile devices to users offers many advantages and disadvantages. Responses about the benefits of lending mobile devices included the following:

- "We love being able to bundle up curated content, and convert non-users. There is nothing like helping someone with a visual issue, such as macular degeneration, learn how easy it is to read on an e-reader. It re-opens the world of literature."
- "Mobile devices are important to bridge the digital divide. More and more titles are e-only, and without lending devices there is no way for some users to access that content. Libraries are about providing access to information/ content."
- "Lending an e-reader to students who do not own or cannot afford one, so that they can better access our e-books."
- "It exposed students, staff, and faculty to the uses of mobile technology."
- "These inexpensive devices will also help us promote our e-collections."
- "Patrons have another option when checking out materials, and they are exposed to different types of technology."
- "It is a break away from the strictly 'book' branding, and allows libraries to have a reputation as a home for cool new technology where faculty and students can try them out."
- "It helps students directly on certain projects where they might not otherwise be able to afford the basic tools needed to complete tasks and/or the tools available from academic departments but only to departmental majors."
- "Lending mobile devices keeps developing staff expertise in new technologies and applications."[10]

However, along with the many benefits added by lending mobile devices there are drawbacks. These were mentioned:

- "Occasional theft from the student and breakage."
- "Annoying losses of cords, chargers, or if a student locks an iPad to an individual account."
- "Hassles purchasing software and book content with university credit cards. Especially doing it on the fly at the request of a user."
- "Difficulty cataloging software and content that are on these devices."
- "Raising awareness on part of students and faculty."
- "Maintenance of the devices when they are returned. We spend a good deal of time making sure all previous patron activity on the device is cleared before we allow it to go back out in circulation. We are considering using a vendor to help with this process, but that is an additional cost we had not originally budgeted for."
- "Storage of user materials on devices and the need to do a factory reset each time one is returned."
- "Fragility of devices."
- "Limited e-book availability in theological studies."[11]

Despite the drawbacks and challenges that lending mobile devices cause, libraries are aware of the importance of this technology in our society and the importance it will continue to play. As a result, more and more libraries are looking to integrate lending mobile devices into their services so that their users can benefit from and be exposed to different types of relevant technology.

ADDITIONAL LIBRARY SERVICES

Mobile technology has also afforded libraries the opportunity to offer additional library services to their users and enhance traditional services such as reference. Text messaging is an example of an additional service that libraries are offering that has been made possible by the development of mobile technology. All that is required for users to take advantage of this service is a mobile device with texting capability. It does not require the user to have an advanced mobile device like a smartphone or tablet. Text message alerts enable libraries to

- notify users of fines and fees that are associated with their accounts
- notify users when holds are ready and available to be picked up
- notify users when materials need to be renewed or returned
- notify users when items that they have checked out are overdue

Signing up to receive these alerts is an easy process. Users are often directed to a website or given a set of instructions with details on how to sign up for text message alerts. This often entails sending a specific message to a phone number or code to sign up and verifying that the message has come through on their mobile device. After this process has been completed, the users is able to receive text message alerts from their library.

Libraries are utilizing text messaging through a service called Text a Librarian. There are a few vendors (Mosio, LibraryH3lp, and Springshare) that provide libraries with this type of service with the need for a mobile device to be used by the library staff. With the service users text a question to a specified number, either a telephone number or a short string of digits, and that question is sent to a dashboard that library staff can log in to and then respond to the question without the need for a mobile device. The top competencies that librarians identified as essential to providing a service of this nature were

- the ability to compose answers to patrons' questions concisely, quickly, and accurately
- the ability to construct effective search strategies and skillfully search online information sources
- the ability to quickly evaluate information and determine the validity, credibility, and authoritativeness of sources
- knowledge of information resources, especially online information resources
- the ability to interpret patrons' information need with the limited context provided in brief text messages[12]

Some library online catalogs include a feature that enables users to send a text message to their mobile phones that have the call number, location, and title of an item. This allows users to navigate to the stacks with information needed to locate an item without the need for paper and pencil. Libraries that have enacted text messaging in some form are Oregon State University Libraries, Eastern Kentucky University Libraries, North Carolina State University Libraries, Indianapolis Public Library, and the Public Library of Cincinnati and Hamilton County, among others.

Mobile technology has enabled libraries to enhance traditional services such as reference. In a time when libraries are facing a decline in subject-related questions at the reference desk, mobile devices such as tablets have allowed reference staff to be more mobile in providing research- and subject-based assistance to users through roving reference. Roving reference is a service that extends beyond the traditional reference desk with the goal of removing barriers between library staff and users. It can be used as an extension of staffing the reference desk or as an alternative way to provide reference services to users. For an academic library, this may mean looking at other venues on campus and outside the library to provide reference services, such as the student center or in the lobby of dorms. Other approaches include having library staff walk throughout the library with tablets to see if users need assistance. This allows library staff to reach and interact with users who need assistance but may not approach the reference desk, and it has the added benefit of making the library staff more visible to users. Additionally, roving reference could be used as a way to enhance and redefine information literacy instruction. Library staff can download various applications and create shortcuts to websites on their devices, which allow them to access the resources with the quick tap of an icon. As mobile technology continues to expand, libraries will continue to look at ways that mobile technology can be integrated to enhance and extend services to benefit users.

NOTES

1. Pew Internet Project. "Mobile Technology Fact Sheet." Pew Research Center, January 2014. http://www.pewinternet.org/fact-sheets/mobile-technology-fact-sheet/
2. Ibid.
3. Johnson, L., S. Adams Becker, V. Estrada, and A. Freeman. *NMC Horizon Report: 2014 Library Edition*. Austin, TX: New Media Consortium. http://cdn.nmc.org/media/2014-nmc-horizon-report-library-EN.pdf
4. Ibid.
5. Ibid.
6. Denny, Heather. "Survey Snapshot: Library Research Using Mobile Devices." MIT Libraries, December 3, 2012. http://libraries.mit.edu/news/survey-snapshot-library/9911/
7. Kim, B. 2013. "The Present and Future of the Library Mobile Experience." *Library Technology Reports* 49, no. 6 (August/September): 15–28.
8. Johnson et al., "NMC Horizon Report."
9. E-mail correspondence with Eryn Roles (Marshall University Libraries), Jonathan LeBreton (Temple University Libraries), Jill McConnell (Cooper-Siegel Community Library), Matthew Collins (Louisville Presbyterian Theological Seminary), and Sarah Felkar (West Vancouver Memorial Library).
10. Ibid.
11. Ibid.
12. Luo, Lili. 2013. "Text a Librarian: Ideas for Best Practices." In *The Handheld Library: Mobile Technology and the Librarian*, edited by Thomas A. Peters and Lori Bell, 43–54. Santa Barbara, CA: Libraries Unlimited.

Chapter Four

Social Media, Mobile Technology, and Libraries

Like mobile technology, social media continues to expand into our everyday lives and has become an essential function of how we use our mobile devices. While it is true that social media is not a purely mobile phenomenon, more and more people are accessing social media platforms like Facebook, Twitter, and Instagram from their mobile devices. In fact, some social media platforms are exclusively built for mobile, like Instagram, Vine, and Snapchat. While you can access Instagram via a desktop or mobile browser, in order to post and apply filters to your images, you need to have the Instagram application installed on your smartphone (the app is currently built for smartphones not tablets). To see the importance of social media and mobile, consider Figure 4.1. In this graphic we can see that mobile technology accounts for a large part of access and traffic to social media platforms like Facebook, Twitter, Instagram, and Pinterest. According to the "Adobe 2014 Mobile Consumer Survey," the top social networking sites frequented by mobile device owners were Facebook (75%), Google+ (29%), Twitter (25%), Instagram (16%), Pinterest (11%), and Snapchat (8%), which gained a significant foothold among younger individuals.[1] With the increasing adoption of mobile technology, specifically smartphones, we are more connected than ever before and have near constant access to information as well as our favorite social media platforms. The emergence of both mobile technology and social media has presented libraries with opportunities to use these additional platforms to reach users.

LIBRARIES AND SOCIAL MEDIA

Social media is a valuable marketing tool for libraries, as well as other organizations, but it is much more than that. Social media enables libraries to reach

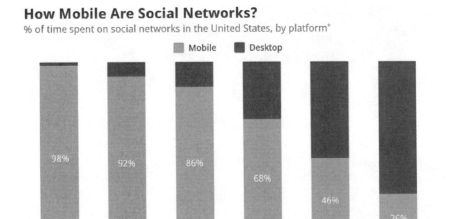

Figure 4.1.   Social Networks and Mobile. *comSource / Statista.com*

users in a digital format that is relatively new, intimidating, and exciting. As with mobile technology, social media is making us as librarians and libraries reevaluate some ways in which we interact with our users and identify how to effectively communicate with them. Determining a good starting point to use social media in your library can seem cumbersome since there are a wide variety of options. Do you go with one, two, three, or more social media platforms? The best answer to this question is to determine what social networks your users are engaging with and focus on those. You may need to start small. Another issue you may face is that your outcomes and interactions are not meeting your expectations. Fortunately, there are many places that you can turn to for help in developing and implementing social media successfully.

In her book *The Librarian's Nitty-Gritty Guide to Social Media*, Laura Solomon offers some valuable insight and advice on how libraries can successfully use social media. She provides insight on how libraries can get started using social media and how they can leverage social media with an understanding of social capital. Social capital, according to Solomon, is what allows any organization or individual to make requests of its followers successfully.[2] This capital allows you to have credibility in an online community. Social capital can be earned over time by participating consistently and appropriately in an online community. Solomon offers ways in which libraries can earn social capital through appropriate uses of social media. Ways to earn social capital include thanking users for engaging and contributing, asking for opinions, offering links to other sites of interest, retweeting your followers, always giving credit for content, encouraging feedback, providing relevant information that your users care about, and monitoring and responding to posts.[3] These are some ways that you and your library can earn social capital and build a positive online reputation.

In a presentation given at the 2015 Computers in Libraries Conference, Cheryl Ann Peltier-Davis, the digital initiatives, cataloging, and metadata services librarian at the University of the West Indies, gave some strategies for success for libraries engaging users with social media. Two important strategies she mentioned were to know why you are using social media tools and to set social media campaign goals. It is important that your library's use of social media provides value through the use of high-quality content. Interaction is essential. Build relationships and connect with your users. Network with industry players at the forefront of developing tools and applications. Lastly, stay informed and keep up-to-date with current trends and changes in the social media landscape.[4]

One important thing that we learned at the Ensor Learning Resource Center is that a social media plan is essential. We had social media accounts with Facebook, Twitter, and Instagram but did not post to them consistently. In order to earn that social capital that we were seeking, we knew we needed to have a more consistent presence on these platforms. As a result, we developed a strategy and a hashtag for each weekday. For Monday, we do #MemeMonday or #ManicMonday, where we try to post some library humor. Tuesday is #TidbitTuesday, which focuses on providing our users with a piece of useful information that they can use, for example, highlighting a database and reminding users to always cite their sources. Wednesday is #WittyWednesday, where we post a picture with emoji and have our followers guess what book is represented by the emoji. Thursday is reserved for #ThrowbackThursday, or #tbt. We try to share something from our special collections and archives on this day. Friday is #FacetimeFriday where we highlight library personnel so our campus can become more familiar with them. This plan has helped us to increase our followers and engagements on all our social media channels. Along with our plan, we also post about events and other important information so our users can feel more connected to us.

USING SOCIAL MEDIA IN THE LIBRARY

Libraries are realizing the value of social media and seizing the opportunities that the various rapidly growing social networks provide to reach users. Here are some examples of ways that libraries can and are using some of the different social media platforms.

Facebook

Facebook was launched in 2004 by then Harvard student Mark Zuckerberg. While it was initially designed as a social networking tool for Harvard students, it quickly spread to others, with more than 1.44 billion global monthly active users as of the first quarter of 2015.[5] Although Facebook was initially launched with desktop users primarily in mind, the company has developed a strong mobile presence. Users can access the platform via a mobile browser or a mobile application that is available on iOS, Android, and Windows Phone. Facebook has a strong presence across a wide range of age groups. Despite rumblings that the younger generation is leaving Facebook for other social networks, it currently remains the most popular social network (Figure 4.2), but that could change. Each social media platform is different, and the ways that libraries and other users approach them are different. Strategies vary based on what social media platform is being utilized.

Libraries are seizing the opportunity to use this large platform to reach users of all ages. Libraries can use Facebook a variety of different ways. Among those are

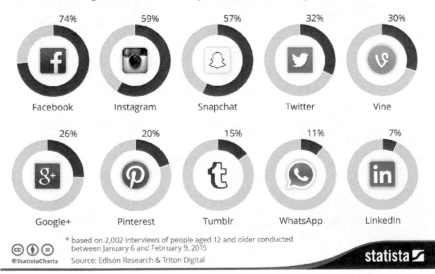

Young Americans Haven't Abandoned Facebook (Yet)
% of Americans aged 12-24 who currently ever use the following social networks*

| 74% | 59% | 57% | 32% | 30% |
| Facebook | Instagram | Snapchat | Twitter | Vine |

| 26% | 20% | 15% | 11% | 7% |
| Google+ | Pinterest | Tumblr | WhatsApp | LinkedIn |

* based on 2,002 interviews of people aged 12 and older conducted between January 6 and February 9, 2015
Source: Edison Research & Triton Digital

statista

Figure 4.2. Facebook and Young Americans. *Edison Research & Triton Digital / Statista.com*

- sharing images and resources from special collections
- promoting and highlighting new resources and services
- posting flyers for library related events
- introducing library staff to users
- posting updates about library hours, programs, or renovations
- posting library related humor and memes

Additionally, The Next Web offers 12 different tactics that organizations, libraries included, can add to their Facebook marketing toolkit to be more effective and engaging. Among them are

- post native videos
- share quote photos
- target your organic posts
- engage your Call to Action button
- try dark posts
- zero in on your key topics
- up your posting frequency
- get creative with trending topics[6]

In 2010, Columbus Metropolitan Library (CML) began a social media campaign to support a levy that would be asking for an increase from voters. It was an unfortunate economic time to be asking for an increase, but their 10-year levy was expiring.[7] The Keep CML Strong Facebook page was created. While Facebook was only one component of the social media campaign, and the awareness campaign in general, it allowed the library to directly interact with users and voters. While it is possible that the levy would have succeeded without the use of social media, the page did garner 3,000 likes and enabled CML to gain some grassroots support.

These are just some examples of ways that libraries can utilize Facebook. Libraries that currently maintain a presence on Facebook include Lexington Public Library, Topeka and Shawnee County Public Library, University of Central Florida Library, Yale University Library, and the Ensor Learning Resource Center, among many others.

Twitter

Founded in 2006, Twitter is a social networking services that allows users to send short 140-character messages referred to as tweets. Twitter is one of the leading social networks worldwide based on active monthly users. As of the first quarter of 2015, Twitter had 302 million active users.[8] Like Facebook,

Twitter is accessible via a desktop or mobile device, but a majority of traffic for Twitter comes via mobile devices in the form of a mobile browser and a mobile application available across the major mobile operating systems (Figure 4.3).

More and more libraries are seeing the value of Twitter and how it can be utilized to connect with users. Twitter can be used in a library setting to

- tweet archival images of students and past events on #tbt
- answer questions about directional information
- highlight and promote library services and resources
- make users aware of library events
- seek feedback from users

It can be challenging to engage your users on Twitter, as it can on any social network. To help with this, Ned Potter offered some insight in an article in the *Library Journal* titled "10 Golden Rules to Take Your Library's Twitter Account to the Next Level." Here are the 10 Golden Rules:

1. Only tweet about your library one time in four.
2. Analyze your tweets.

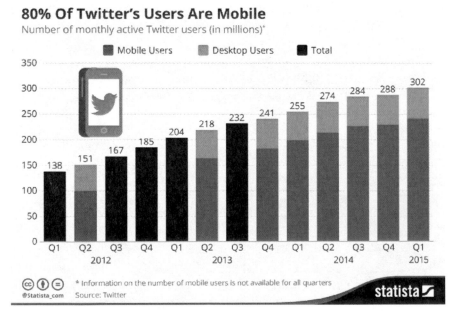

Figure 4.3.   Mobile Users and Twitter. *Twitter / Statista.com*

3. Tweet multimedia.
4. Tweet more pictures.
5. If something is important, tweet it four times.
6. Use hashtags (but don't go mad).
7. Ask questions.
8. Get retweeted and your network will grow.
9. Put your Twitter handle everywhere.
10. Avoid pitfalls like using posting pictures from apps that don't display the images on Twitter.[9]

A specific example of how libraries have used Twitter is the University of Buffalo Libraries. In May 2015, they posted an announcement to let users know that could stop by their library for therapy dogs and coffee/snacks to help them deal with potential stress that the end of the semester was bringing. The tweet read "Take a break and relax with therapy dogs during Stress Relief Days on South Campus." It included an image of dogs and a link to a blog post for more info.

Twitter can be used is a variety of different ways; these are just some examples of ways that libraries can use this platform. Libraries that are using Twitter include Binghamton University Libraries, Miami University Libraries, Stony Brook University Libraries, Kenton County Public Library, New York Public Library, and the Clinton Presidential Library. Twitter is a valuable tool that lets you interact with users with a short amount of characters and is becoming increasing popular with libraries.

Instagram

Instagram is a photo-sharing social network that was launched in 2010. Facebook acquired Instagram for $1 billion in cash and stocks in April 2012. As of December 2014, Instagram had more than 300 million active users.[10] Unlike both Facebook and Twitter, Instagram is specifically designed for mobile devices, more specifically smartphones. In order to use Instagram, you need to have the application installed on your smartphone. The app is available on the iOS, Android, and Windows Phone platforms.

Instagram is a great tool that allows libraries to connect to users with images. Ways that libraries can use Instagram include

* highlighting new additions to the collection
* posting images from special collections
* advertising library events
* creating a photo contest

- posting images of users in your library
- introducing and promote library staff
- highlighting useful resources and services

In May 2015, Bernardsville Public Library created a campaign on Instagram called #libraryinmyhand (Figure 4.4). The goal of this initiative is to promote mobile access to resources available at the library. They have edited screenshots and are using the slogan "We put the library in your hand!" Other libraries have followed suit and have started to use #libraryinmyhand to promote access to mobile resources to users.

Libraries are using Instagram in a variety of different ways to connect with users. Libraries that are using Instagram include University of Oregon Libraries, Central Michigan University Libraries, Boone County Public Library, Cleveland Public Library, and Toronto Public Library.

Pinterest

Launched in 2010, Pinterest is a content-sharing service that allows members to "pin" images, videos, and other objects to their pinboard. It became the fastest site in history to reach 10 million unique monthly visitors and high user engagement metrics even though during its launch it was accessible by invitation only. It has a strong female presence and is one of the world's top referral traffic generating website online. As of January 2015, there were 50.5 million Pinterest users in the United States.[11] It saw a dramatic increase in unique visitors from 2011 to 2015 (Figure 4.5). Pinterest is available across a wide range of platforms. It is accessible on a computer, a mobile browser, and a mobile application available on iOS, Android, and Windows Phone.

Given the pinboard nature of Pinterest, libraries can use this social media platform for a host of different things. According to the American Library Association, libraries have been using Pinterest in the following ways:

- Pinning book covers. Many librarians use the visual power of Pinterest to display book covers, especially those from new books, special collections, and child-friendly material.
- Creating reading lists on a wide range of topics.
- Getting the word out on recent acquisitions.
- Fostering research. A lot of Pinterest material is on the light side, but some librarians and academics see potential in the site for much more serious applications.
- Promoting library activities, showcasing everything from lectures to job help and author visits.

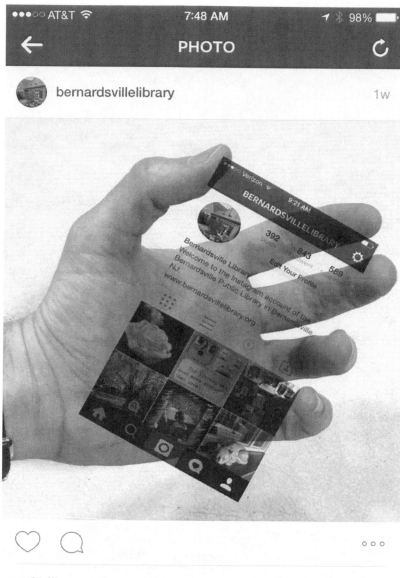

bernardsvillelibrary   1w

♡ ◯   ○ ○ ○

❤ **91 likes**

bernardsvillelibrary @bernardsvillelibrary STARTED A NEW TREND FOR LIBRARY PR AND TAGGED IT #libraryinmyhand. Here's the FIRST ever #libraryinmyhand pic, which is ours, of course! The #libraryinmyhand idea is based on the #instainmyhand pics popular in #Japan. I've

Figure 4.4.   Bernardsville Public Library #libraryinmyhand

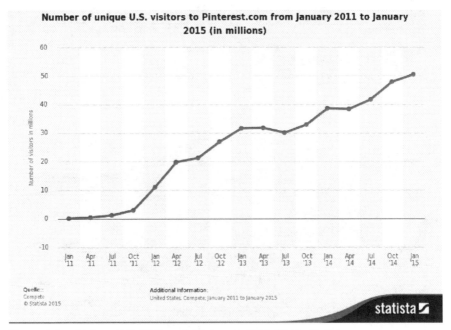

Figure 4.5.   Unique U.S. Visitors to Pinterest, 2011–2015. *Statista.com*

- Offering access to digital collections. With e-books gaining popularity, some libraries are using Pinterest to share links to new digital materials.[12]

   The University of Texas (UT) Libraries are one library system that is using Pinterest. They have created a guide explaining why they like to use Pinterest for their library. Those reasons are

- Ease of Use—Setting up and maintaining the site are both easy tasks. Additionally, you can maintain a strong presence on Pinterest without updating the site every day.
- Outreach—You can connect with other users across social media platforms via Pinterest by quickly pinning and linking to content on other sites. On the UT Pinterest boards, we pin content from Instagram taken by our core user base, students using the libraries.
- Promotion—Pinterest can be a great way to highlight tools, resources, or aspects of the library that can get lost in the shuffle.
- Collaboration—Aside from reaching out to our users (such as students), Pinterest can be a great tool for collaboration and a way to either connect with or work with other organizations on campus.

- Curation—Pinterest is a space for digital curation and librarians can draw upon their expertise in curating physical objects in the virtual space of Pinterest.
- Visualizing Value—Pinterest can be a way to subvert expectations and reimagine the academic library. Pinterest can be used to promote modern aspects of the academic library that are unique and even unexpected. As a tool, Pinterest lends itself to sharing powerful visuals, new technologies, humor, events, ideas, and aspects of academic libraries that appeal to our user base, such as digital collections.[13]

Libraries are using Pinterest in some very interesting ways, and some libraries have even used Pinterest as an inspiration for their websites. Libraries that are currently on Pinterest include Virginia Tech University Libraries, University of Texas Libraries, San Francisco Public Library, and Fullerton Public Library.

Snapchat

Snapchat is a photo- and video-sharing app that allows users to send images and videos, referred to as snaps, to a controlled list of recipients; the snaps self-destruct after a short amount of time (usually between 1 and 10 seconds). It was created by Evan Spiegel, Bobby Murphy, and Reggie Brown while they were attending Stanford University in 2011. Facebook offered to buy Snapchat for $3 billion, but the offer was turned down. Snapchat, like Instagram, was designed for mobile, specifically smartphones. It is available as a mobile application on iOS and Android.

Snapchat is an intriguing social media tool for libraries. According to Figure 4.6, it is more popular among American millennials than Twitter. Of particular interest for academic libraries is that 77% of college students say that they use Snapchat on a daily basis.[14] With this many potential library users actively utilizing Snapchat, it is worth looking at ways that it can be used to connect with them. Some ways that Snapchat can be used in libraries are

- creating a Snapchat library video contest
- promoting library services and resources
- introducing library staff
- sharing short information videos

One way that libraries can start implementing Snapchat for promotional purposes is through the use of Snapchat Stories. This feature enables Snapchat users to take a series of snaps and select them to "Snap to Your Story."

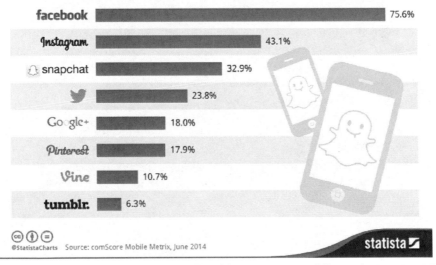

Snapchat More Popular Than Twitter Among Millennials

Most popular social media apps among Americans aged 18-34 (% of smartphone users)

facebook — 75.6%
Instagram — 43.1%
snapchat — 32.9%
(twitter) — 23.8%
Google+ — 18.0%
Pinterest — 17.9%
Vine — 10.7%
tumblr. — 6.3%

@StatistaCharts   Source: comScore Mobile Metrix, June 2014   statista

Figure 4.6.   Snapchat and American Millennials. *comSource Mobile Matrix, June 2014 / Statista.com*

The photos in your story are available for viewing by all of your Snapchat friends for 24 hours. The *Librarian Enumerations* blog offers some ideas on how libraries can utilize this feature of Snapchat. Some story ideas include

- a day in the life of a librarian
- how books make it to the shelf
- how digital records are searchable (the process of metadata)
- steps in the research process[15]

   One library that has been using Snapchat is the University of Mary Hardin-Baylor Library.

CONCLUSION

In addition to those mentioned, there are a variety of other social media platforms—such as Tumblr, Google+, YouTube, and WhatsApp—with a strong mobile presence that libraries are taking advantage of in order to reach and engage users. While social media is not exclusively mobile, many of the newer social networks are solely focused on mobile and others maintain a strong mobile presence. The challenge for us and our libraries is to choose the social media platforms that we believe will effectively reach and engage our users.

NOTES

1. Adobe. "Adobe 2014 Mobile Consumer Survey." White Paper, 2014.

2. Solomon, Laura. *The Librarian's Nitty-Gritty Guide to Social Media.* Chicago: American Library Association, 2013.

3. Ibid.

4. Peltier-Davis, Cheryl Ann. "Syncing for Success: Social Media & Mobile Apps Tips & Tools for Innovative Services." Presentation at the Computers in Libraries Conference, Washington, DC, April 29, 2015. http://conferences.infotoday.com/documents/219/A302_Peltier-Davis.pptx

5. Statista. "Facebook—Statistics & Facts." http://www.statista.com/topics/751/facebook/

6. Seiter, Courtney. "12 Latest Tactics for Posting on Facebook." The Next Web, May 1, 2015. http://thenextweb.com/socialmedia/2015/05/01/12-latest-tactics-to-posting-on-facebook/

7. Circle, Alison. "Marketing a Levy Through Social Media." *Library Journal*, January 19, 2011. http://lj.libraryjournal.com/2011/01/opinion/bubble-room/marketing-a-levy-through-social-media/

8. Statista. "Twitter—Statistics & Facts." http://www.statista.com/topics/737/twitter/

9. Potter, Ned. "10 Golden Rules to Take Your Library's Twitter Account to the Next Level." Library Journal. August 27, 2013. http://lj.libraryjournal.com/2013/08/marketing/10-golden-rules-to-take-your-librarys-twitter-account-to-the-next-level/#_

10. Statista. "Instagram—Statistics & Facts." http://www.statista.com/topics/1882/instagram/

11. Statista. "Pinterest—Statistics & Facts." http://www.statista.com/topics/1267/pinterest/

12. American Library Association. "Social Networking." 2013. http://www.ala.org/news/state-americas-libraries-report-2013/social-networking_

13. University of Texas Libraries. "Course Guides: Visualizing Value: Using Pinterest to Market the Academic Library." University of Texas Libraries LibGuide. June 19, 2014. http://guides.lib.utexas.edu/subjects/guide.php?subject=pinterest

14. Wagner, Kurt. "Study Finds 77% of College Students Use Snapchat Daily." Mashable, February 24, 2014. http://mashable.com/2014/02/24/snapchat-study-college-students/

15. Alfonzo, Paige. "Promote Your Library with Snapchat Stories." *Librarian Enumerations* (blog), June 27, 2014. https://librarianenumerations.wordpress.com/2014/06/27/promote-your-library-with-snapchat-stories/

Chapter Five

Responsive Web Design

With the increase in mobile web traffic and continued development of feature-rich mobile devices, end users expect to be able to consume information that is formatted and responds to the device of their choice, be it a desktop, tablet, and/or smartphone. The advancement in mobile technology and increasing expectations from users have led to a shift in the way that websites are designed. Instead of focusing on building websites formatted for fixed-width displays—such as desktops, which are typically 1024px—web designers and developers are starting to take advantage of the fluid and flexible nature of the web to create better viewing experiences across a broad range of screen sizes. Websites have been designed similar to the print medium, with the idea of a fixed-width display in mind. In fact, in 2000 designer Jon Allsopp pointed out that the web does not have the same constraints as print. He stated that

> the control which designers know in the print medium, and often desire in the web medium, is simply a function of the limitation of the printed page. We should embrace the fact that the web doesn't have the same constraints, and design for this flexibility. But first, we must accept the ebb and flow of things.[1]

This flexibility, along with the power of mobile, has businesses and organizations rethinking their approach to mobile by striving to meet user expectations by providing them with optimal viewing experiences across a wide range of devices and screen sizes. Given that libraries provide users with access to a wealth of information and that they are spending an increasing amount on digital content, it stands to reason that libraries are just as concerned as other businesses about meeting user expectations by providing mobile-optimized access to resources. The proliferation and diversity of screen sizes has led to a shift in web design where a mobile-first approach is being taken by utilizing responsive web design.

WHAT IS RESPONSIVE WEB DESIGN?

Put simply, responsive web design is a process of developing websites that aim to create an optimal viewing experience across a wide range of devices from smartphones to desktops. The term was made popular by Ethan Marcotte, a web designer and developer, in 2010. He stated that responsive web design offers us a way forward, finally allowing us to "design for the ebb and flow of things."[2] The rising popularity of this approach to web design can most certainly be attributed to the diversity of screen sizes brought about by the mass adoption of mobile devices. However, in order to fully understand what responsive web design is, it is important to know the differences between responsive web design, mobile websites, and adaptive web design.

Responsive Web Design and Mobile Websites

It may seem logical to think that since responsive web design takes a mobile-first approach, it means that you are creating a mobile website. However, that is not the case. Developing a website using responsive web design differs from creating a mobile website in many ways. A mobile website is a separately developed website. It has a different URL, such as http://yourlibrary.org/m, that is different from the main library website URL (such as http://yourlibrary.org). The mobile website tends to be a scaled down version of your main site, with information that your library has deemed important for users or something that mobile users want to access on their devices. On the other hand, responsive websites deliver the same content to your users using the same URL on any device. Responsive web design provides a website that is scalable and responds to the device to fit the needs of the user and provides a viewing experience formatted for that device. This is done using CSS and HTML.

With mobile devices becoming more useful tools for productivity, we can no longer make assumptions about what our users want and water down our library websites. User expectations have shifted. No longer are people using mobile devices for just idle web browsing; they are using them for a variety of purposes. As a result, libraries need to ensure that our websites reflect this changing nature and show our users that we care about and understand their information access needs. If mobile device users visit a website that is not mobile optimized, they are less likely to remain on that website (speaking from personal experience). Responsive web design is Google's recommendation for building websites as it helps eliminate some of the common mistakes of mobile websites. The common mistakes of mobile websites that Google lists are

- blocked JavaScript, CSS, and image files
- unplayable content

- faulty redirects
- mobile-only 404s
- avoid interstitials
- irrelevant cross-links
- slow mobile pages[3]

Additionally, mobile websites have to be maintained separate from your main website. As a result, when content changes on your main website, it may also need to be changed on your mobile website. These are some of the distinct differences between responsive web design and mobile websites.

Responsive Web Design and Adaptive Web Design

A competing way to design and develop websites for a diverse range of devices and screen sizes is adaptive web design. While both responsive web design and adaptive web design aim to scale websites down to display and format nicely on different devices, there are some distinct differences between the two approaches.

Adaptive web design uses static layout templates based on various breakpoints and detects the screen size of your device to determine which layout to display. This approach uses the server to detect the screen size. The Mozilla Developer Network, an open community of developers that provides resources for web development, defines adaptive web design as follows:

> Adaptive design is more like the modern definition of progressive enhancement. Instead of one flexible design, adaptive design detects the device and other features, and then provides the appropriate feature and layout based on a predefined set of viewport sizes and other characteristics. This can result in a lack of consistency across platforms and devices, but the load time tends to be faster.[4]

The reason that adaptive web design tends to load fast is that it only loads the template needed based on the screen size of the device; the other templates are not loaded.

On the other hand, responsive web design utilizes a specific CSS code (CSS media queries) to modify the presentation of the website based on the device it is being viewed on. The Mozilla Developer Network defines responsive web design as follows:

> Responsive design works on the principle of flexibility. The idea is that a single fluid design based upon media queries, flexible grids, and responsive images can be used to create a user experience that flexes and changes based on a multitude of factors. The primary benefit is that each user experiences a consistent design. One drawback is a slower load time.[5]

Responsive web design has slower load times than websites using adaptive web design because information regarding each device is being download once the website loads, regardless of whether that code is currently being utilized.

There are some other distinct differences between the two approaches that you should be aware of. Adaptive web design requires you to maintain separate websites either by URL or by separate HTML and CSS code, while responsive web design relies on existing HTML, CSS, and JavaScript so you don't have to maintain separate websites. Adaptive web design relies on predefined screen sizes for layouts (e.g., 320, 480, 768, etc.), while responsive web design relies on flexible and fluid grids. Adaptive web design tends to be easier to implement on existing websites, while many newer websites tend to use responsive web design. Adaptive web design can require a more advanced knowledge of JavaScript, while responsive web design relies on HTML and CSS. Lastly, adaptive web design uses images that are optimized for specific screen resolutions, while responsive web design contains images that are first downloaded and re-sized to fit the screen.

Elements of Responsive Web Design

Before responsive web design was introduced, many web developers, including libraries, tended to follow the principle of pixel-perfect web design. Pixel-perfect web design applies the same process to a web page as many magazine publishers, where there is a fixed width for content and a mock-up is often done in a graphics program like Adobe Photoshop. The goal of pixel-perfect web design is to make the finished product look as much like the mock-up as possible.[6] The problem with this approach is that websites that are designed with fixed-width layout are not scalable and provide a poor viewing experience on smaller devices such as smartphones. This approach also defines elements on web pages to be specific pixel-based widths. An example of this would be the CSS markup for a container on a web page:

```
.container {
        width: 960px;
}
```

This approach is fine for websites that are view on desktops, but does not work well on smaller devices. As a result, users have to pinch and zoom, which creates a frustrating user experience. The rise in mobile web traffic has given rise to responsive web design.

Responsive web design aims to solve the issues of pixel-perfect web design through the use of three essential elements:

- flexible, grid-based layout
- media queries
- fluid images

Flexible grids are created using relative units of measurement, like percentages, instead of absolute units of measurement, like pixels. Take, for instance, the previous example; instead of defining the container width in pixels, we can define it with a percentage so it become something like this:

```
.container {
        width: 95%;
}
```

Using relative units of measurements to create a grid-based layout on your website allows for greater flexibility because the content will reformat to the percentage regardless of the screen size. Using a flexible grid, also referred to as a fluid layout, you design the maximum layout for your website and divide your grid into a specific number of columns. Designing the elements with proportional widths and heights will enable them to readjust across different devices and screen sizes. There are a variety of different responsive grid-based systems available on the web, or you can create your own fluid layout using CSS. Most of these grid systems are based on a width of 960px with a 12-column grid (since 960 divides equally by 12).

Media queries are another essential element of responsive web design. Media queries allow you to add different CSS styles to elements on your website depending on screen size and width. This allows for elements to respond differently based on the device that is being used to access a website. These can be applied all within one CSS file by using the syntax similar to the following:

```
@media screen and (min-width: 900px) {…}
@media screen and (max-width: 768px) {…}
@media screen and (max-width: 480px) {…}
```

We can add CSS within each one of these media queries to have elements on our website render differently based upon the minimum or maximum size of the screen. For example, the min-width property in the media query will apply styles that we have defined with the curly braces to websites being viewed on devices where the widths are at least 900px, and the max-width property will apply certain styles based on devices that have a max width of 768px. Any device above the max-width will have the normal styles applied to it. It is important to note that the min-width and max-width values in this example are used for the purpose of explanation. You may determine that your library website needs to have different breakpoints to properly adjust for various devices. The great thing about media queries is that they enable you to apply certain styles based upon the width of devices users may be using to access your website. You won't need to redefine all elements on your web page using media queries, but there will be certain elements that you will need to rearrange on your website to accommodate various devices. For example, say that your website has three boxes that span the width of the page when viewed on a desktop computer. You will need to write some CSS in a media query to change the display of those boxes to accommodate smaller screen sizes. Let's say your normal CSS width for those boxes looks something like

```
.boxes {
        width: 33.333%;
}
```

You may decide that you want those boxes to span the entire width of the page once the width of a device is 768px. You would add the following to your CSS file:

```
@media screen and (max-width: 768px) {
    .boxes {
            width: 100%;
    }
}
```

If the only thing that you are needing to change is the width of the box; that is the only part that needs to go in the media query. Other styles like background color, font color, or border radius will still be applied to the box because the only thing that you are telling the media query to change is the width of the boxes.

Since responsive web design takes into account various different screen sizes, there is the need to ensure that the images on a website respond and scale to the device that users are viewing a website on. Using CSS you can make images more flexible and fluid by controlling them and other fixed-width elements to ensure that they correspond to the width of the container that they are assigned to. To make your images fluid and ensure that they don't bust out of their assigned containers, you can add a max-width to your images like the following:

```
img {
      max-width: 100%;
}
```

This will ensure that your images are fluid and free to scale up or down in relation to the container they are housed in.

These three elements make up the core of the responsive web design approach. Due to the increasing adoption of this approach to developing websites, there are several responsive frameworks that are available and content management system (CMS) themes that make it easier to implement a responsive website for your library.

RESPONSIVE FRAMEWORKS

There are different responsive frameworks available to help you build a responsive website for your library. One of those frameworks is Bootstrap. Bootstrap is a free and open source front-end framework using HTML, CSS, and JavaScript that enables developers to build responsive, mobile-first websites. The framework was designed and created by Mark Otto and Jacob Thornton of Twitter to improve some internal tools and create consistency. However, after some development they decided to make it freely available. In an article from 2012, Mark Otto stated:

> A small group of Twitter employees set out to improve our team's internal analytical and administrative tools. After some early meetings around this one product, we set out with a higher ambition to create a toolkit for anyone to use within Twitter, and beyond. Thus, we set out to build a system that would help folks like us build new projects on top of it, and Bootstrap was conceived.[7]

At the outset, Bootstrap only included typography, tables and forms, a color guide, and some graphical assets.[8] However, after more collaboration with developers and designers, more features were added to the framework. As a result, Bootstrap is now one of the most popular responsive frameworks for developing responsive websites.

There are several ways that you can use Bootstrap to develop a responsive website. You can download Bootstrap from GitHub, or you can use the content delivery network (CDN) option to link to Bootstrap. To link to the

Bootstrap CDN files, you will need to include the following links in your HTML document:

```
<!-- Latest compiled and minified CSS -->
<link rel="stylesheet"
href="https://maxcdn.bootstrapcdn.com/bootstrap/3.3.5/
css/bootstrap.min.css">

<!-- Optional theme -->
<link rel="stylesheet" href="https://maxcdn.boot-
strapcdn.com/bootstrap/3.3.5/css/bootstrap-theme.min.
css">

<!-- Latest compiled and minified JavaScript -->
<script src="https://maxcdn.bootstrapcdn.com/boot-
strap/3.3.5/js/bootstrap.min.js"></script>
```

You also want to make sure that you add a link to jQuery file because it is required to use any of Bootstrap's JavaScript extensions.

In addition to providing the source code and CDN links, the Bootstrap website includes a starter template and good documentation with code snippets to build off of the template and other examples. To build off of the starter template, view the source of the template in your browser and copy the code. The code for the starter template should look something like this:

```
<!DOCTYPE html>
<html lang="en">
  <head>
    <meta charset="utf-8">
    <meta http-equiv="X-UA-Compatible" content="IE=
    edge">
    <meta name="viewport" content="width=device-width,
    initial-scale=1">
    <!-- The above 3 meta tags *must* come first
    in the head; any other head content must come
*after* these tags -->
    <meta name="description" content="">
    <meta name="author" content="">
    <link rel="icon" href="../../favicon.ico">

    <title>Starter Template for Bootstrap</title>
```

```
<!-- Latest compiled and minified CSS -->

    <link rel="stylesheet" href="https://maxcdn.
bootstrapcdn.com/bootstrap/3.3.5/css/boot-
strap.min.css">

    <!-- Optional theme -->
    <link rel="stylesheet" href="https://maxcdn.
bootstrapcdn.com/bootstrap/3.3.5/css/boot
    strap-theme.min.css">

    <!-- Custom styles for this template -->
  <link href="http://getbootstrap.com/examples/
  starter-template/starter-template.css" rel=
"stylesheet">

  </head>

  <body>

    <nav class="navbar navbar-inverse navbar-fixed-
top">
  <div class="container">
   <div class="navbar-header">
   <button type="button" class="navbar-toggle col-
lapsed" data-toggle="collapse" data-target="#navbar"
aria-expanded="false" aria-controls="navbar">
    <span class="sr-only">Toggle navigation</span>
    <span class="icon-bar"></span>
    <span class="icon-bar"></span>
    <span class="icon-bar"></span>
   </button>
   <a class="navbar-brand" href="#">Project name
   </div>
  <div id="navbar" class="collapse navbar-collapse">
   <ul class="nav navbar-nav">
    <li class="active"><a href="#">Home</li>
    <li><a href="#about">About</li>
    <li><a href="#contact">Contact</li>
   </ul>
  </div><!--/.nav-collapse -->
```

```
  </div>
</nav>

  <div class="container">

  <div class="starter-template">
  <h1>Bootstrap starter template</h1>
  <p class="lead">Use this document as a way to
quickly start any new project.<br> All you get is
this text and a mostly barebones HTML document.</p>
  </div>

  </div><!-- /.container -->

  <script src="https://ajax.googleapis.com/ajax/
libs/jquery/1.11.3/jquery.min.js"></script>

  <script src="https://maxcdn.bootstrapcdn.com/boot-
strap/3.3.5/js/bootstrap.min.js"></script>
  </body>
</html>
```

This template includes a navigation bar and a container with text. When viewed in a browser, the template should look like Figure 5.1. Using this template, you can explore the other components of the framework, such as themes, grids,

Bootstrap starter template

Use this document as a way to quickly start any new project.
All you get is this text and a mostly barebones HTML document.

Figure 5.1.   Bootstrap Starter Template

and navigation bars, and add your own custom elements to build a responsive website for your library. The additional components of this framework include glyphicons, dropdowns, various button styles, panels, and responsive embed among many other components. Bootstrap is supported by the latest versions of major web browsers like Google Chrome and Firefox. As mentioned previously, this is an extremely popular framework. In fact, the newest version of Springshare's LibGuides is built using Bootstrap (more on that in the next chapter).

Another responsive framework is ZURB Foundation, or just Foundation. Like Bootstrap, Foundation is an open source front-end framework with a collection of tools for creating websites and web applications. It contains HTML- and CSS-designed templates as well as optional JavaScript extensions to add more interactivity to your website. This framework comes packaged with a flexible grid layout, a pre-built CSS file for all of the HTML elements, advanced components, and JavaScript plugins and extensions based on Zepto.js (a lighter alternative to jQuery). The Foundation framework is built with small mobile devices, that is, smartphones, as a main priority. It is built with the Sass (Syntactically Awesome Stylesheets) stylesheet language (Sass is a scripting language that is interpreted into CSS to style the HTML elements). Like Bootstrap, Foundation is a fully customizable framework.

In addition to responsive frameworks, there are several responsive themes if your library uses a CMS like WordPress and Drupal. These themes allow you to create a responsive website for your library by building off of the default template. This eliminates the need to build a responsive website from the ground up. In fact, some of the themes for these CMSs are built using responsive frameworks.

WHY USE RESPONSIVE WEB DESIGN

There are many distinct advantages to using responsive web design to develop library websites, or any website. Responsive web design eliminates the need to develop and maintain a separate mobile website, because all the content on the website will respond based on the screen size of the device. This makes navigation easier from a user's perspective. It also provides a familiar interface and experience across a broad range of devices. Responsive web design is recommended by Google because it

• Makes it easier for users to share and link to your content with a single URL.
• Helps Google's algorithms accurately assign indexing properties to the page rather than needing to signal the existence of corresponding desktop/ mobile pages.

- Requires less engineering time to maintain multiple pages for the same content.
- Reduces the possibility of the common mistakes that affect mobile sites.
- Requires no redirection for users to have a device-optimized view, which reduces load time. Also, user agent-based redirection is error-prone and can degrade your site's user experience.
- Saves resources when Googlebot crawls your site. For responsive web design pages, a single Googlebot user agent only needs to crawl your page once, rather than crawling multiple times with different Googlebot user agents to retrieve all versions of the content. This improvement in crawling efficiency can indirectly help Google index more of your site's content and keep it appropriately fresh.[9]

While we don't need to let Google dictate how we design our websites, on a mobile browser Google will let a user know if a website is mobile friendly (Figure 5.2). This feature was rolled out by Google in November 2014 in an effort to provide users with a better mobile experience by letting them know which sites are formatted for mobile devices. A website is eligible for the mobile-friendly tag if it meets the following criteria as detected by Googlebot:

- avoids software that is not common on mobile devices, like Flash
- uses text that is readable without zooming
- sizes content to the screen so users don't have to scroll horizontally or zoom
- places links far enough apart so that the correct one can be easily tapped[10]

Google has expanded its use of the mobile-friendly criteria as a ranking signal, which makes having a mobile-friendly library website all the more important. The changing of Google's algorithm means that mobile-friendly websites are given preference on searches performed using mobile devices.

To check to see how responsive, or mobile-friendly, your library's website is, you can use several free tools like Responsinator and Google's Mobile-Friendly Test. Responsinator allows you to enter the URL of your website and see how it displays across a broad range of device like an iPhone, iPad, Android smartphone, Android tablet, and Kindle Fire, among others. Google's Mobile-Friendly Test asks for the URL of your website, analyzes the site to determine if it is mobile friendly, and provides you with a screenshot of how Googlebot see the web page.

More and more websites are being developed using responsive web design. Examples of library websites using responsive web design are North Carolina State University Libraries, Eastern Kentucky University Libraries, Berea College Library, Canton Public Library, and the New York Public Library. One

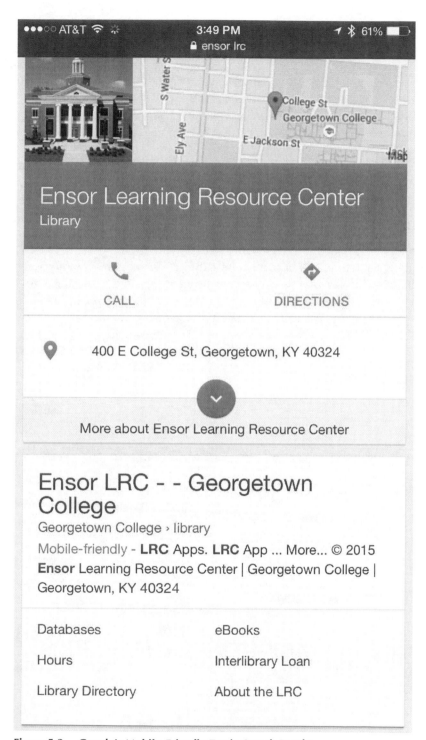

Figure 5.2. Google's Mobile-Friendly Tag in Search Results

reason libraries are moving to a responsive website is that they are finding that users are accessing the full website more and more from their mobile devices and bypassing the mobile website. What this signifies is that users want access to all of the content and information on their mobile device that they would get viewing the library website from a desktop. This was a major reason that my library, the Ensor Learning Resource Center at Georgetown College, decided to undertake a website redesign.

My library initially undertook a website redesign in 2012 when we moved to WordPress. The goal was to enhance our library website and develop a mobile website using jQuery Mobile to accommodate mobile users. Using Google Analytics to track our statistics, we found that 3,674 (or about 5%) of the 70,718 visits to our library website between June 1, 2012 (when the new site was launched) and December 31, 2013, were from mobile devices. During that same time period there were only 483 visits to our mobile website. Using this information, we determined that it would be beneficial for our users to move to a responsive website. Looking through several responsive WordPress themes we decided to use the Responsive Theme by Cyberchimps (Figure 5.3). After a few iterations and customizations of the theme, we launched our new responsive website in tandem with our new integrated library system in May 2014 (Figure 5.4). The header and menu are elements

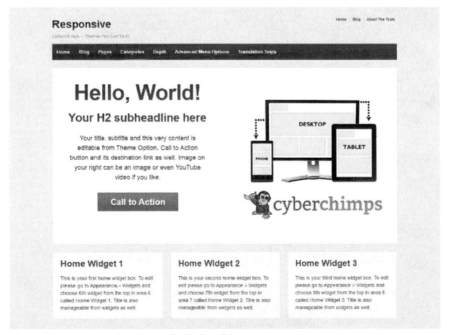

Figure 5.3.   Responsive Theme by Cyberchimps

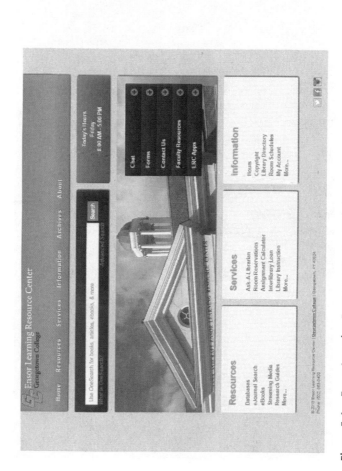

Figure 5.4. Ensor Learning Resource Center Website

that we included from the Bootstrap framework. In the first year after we launched the new website, we had 30,273 visits with 3,952 coming from mobile devices. Mobile device visits in that year accounted for 13% of our total visits. This reaffirmed our decision to move to a responsive website and signaled to us that our users do not want their options limited when they visit our website from their mobile devices.

With the explosion of mobile technology, having a website that responds to a broad range of devices is essential now and moving forward. Responsive websites can help provide a better user experience, and they are the preference for search engine optimization. It is no wonder why responsive web design is considered by some to be a big part of the future of web design.

NOTES

1. Allsop, John. "A Dao of Web Design." *A List Apart*, April 7, 2000. http://alistapart.com/article/dao

2. Marcotte, Ethan. "Responsive Web Design." *A List Apart*, May 25, 2010. http://alistapart.com/article/responsive-web-design

3. Google. "Avoid Common Mistakes." Google Developers, 2015. https://developers.google.com/webmasters/mobile-sites/mobile-seo/common-mistakes/

4. Mozilla Developer Network. "Responsive design versus adaptive design." January 26, 2015. https://developer.mozilla.org/en-US/Apps/Design/UI_layout_basics/Responsive_design_versus_adaptive_design

5. Ibid.

6. Kim, Bohyun. 2013. "The Library Mobile Experience: Practices and User Expectations." *Library Technology Reports* 49, no. 6 (August/September): 29–39.

7. Otto, Mark. "Building Twitter Bootstrap." *A List Apart*, January 17, 2012. http://alistapart.com/article/building-twitter-bootstrap

8. Ibid.

9. Google. "Responsive Web Design." Google Developers, 2015. https://developers.google.com/webmasters/mobile-sites/mobile-seo/configurations/responsive-design

10. Google. "Helping Users Find Mobile-Friendly Pages." *Google: Webmaster Central Blog*, November 18, 2014. http://googlewebmastercentral.blogspot.com/2014/11/helping-users-find-mobile-friendly-pages.html

Chapter Six

What Library Vendors Are Offering

The expanding importance and reliance on mobile technology has caused both libraries and library vendors to look at ways to provide access to content in a mobile-friendly format. With more people accessing information from the smaller mobile device displays, libraries are relying on vendors to provide mobile access to subscribed content. Many vendors have met this expectation and have developed ways for users to access content their library subscribes to through the development of mobile web interfaces or native applications that target specific mobile platforms.

MOBILE WEB INTERFACES

One way that library vendors are providing mobile access to their content is through the development of mobile web interfaces. Vendors who have chosen this means are EBSCO, Springshare, Annual Reviews, Taylor & Francis, BioOne, and Web of Knowledge, among a host of others.

EBSCO

EBSCO Information Services is a company that provides research resources comprised of research databases, e-books, and e-journals to libraries of all types. EBSCO is one library vendor that has provided access to content through mobile web interfaces. Previously EBSCO provided mobile access through a separate mobile profile, mobsmart. While libraries can currently offer mobile access through this profile, EBSCO has developed an enhanced mobile interface. The new interface, dubbed EBSCOhost Mobile 2.0, uses mobile detection technology to route users to the appropriate version of EBSCOhost depending on their device without the need for an additional mobile profile. Features of EBSCOhost Mobile include

- mobile device detection
- basic search and results list
- viewing HTML and PDF full text
- e-mail record with full text
- multiple database searching
- setting user preferences
- guest access
- integration with EBSCO Discovery Services (EDS)

While most of these features are available through either the mobsmart profile or EBSCOhost Mobile 2.0, the latter does offer some additional features such as mobile detection, integration with EDS, guest access, and personal user authentication. EBSCO has sought to offer mobile access through a variety of different means, and it is currently one of the only library vendors that has focused on both native application and mobile web interface development.

Springshare

Springshare was founded in 2007 with the goal of developing web applications for libraries and educational institutions. One of the leading products that Springshare offers for libraries is LibGuides. LibGuides, as well as other Springshare products, has recently undergone a redesign to ensure optimal usability across a broad range of devices. Previous to this redesign, users accessing their library's LibGuides via a mobile device were redirected to a stripped-down mobile site that was usually devoid of any institutional colors and logo. Although the content generally remained the same, the interface was not that attractive. To improve upon the interface, Springshare developed a new version using responsive web design by using the Bootstrap framework. In addition to updating LibGuides, Springshare has also updated the interfaces of their other products such as LibAnswers. With the updated design users are able to access the complete LibGuide with colors and logos, which creates a consistent viewing experience across devices (Figure 6.1).

Annual Reviews

Annual Reviews is a nonprofit scientific publisher whose volumes are published each year for more than 41 focused disciplines within the biomedical, life, physical, and social sciences. They have focused on offering mobile

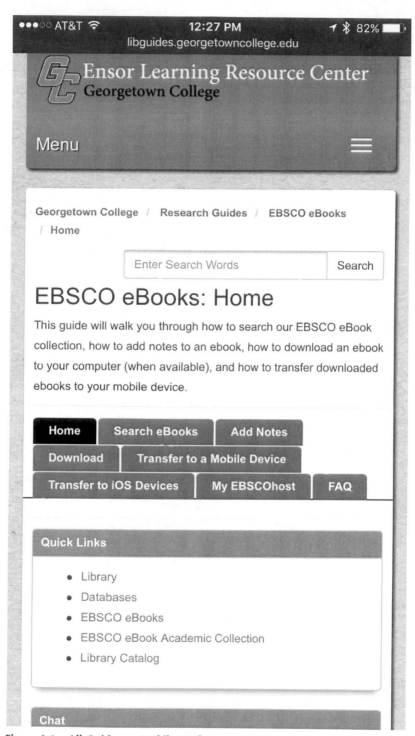

Figure 6.1.   LibGuides on Mobile Device

access through the development of a mobile web interface. Access is available and optimized for most mobile platforms. The Annual Reviews mobile website enables users to

- browse journal subscriptions and articles
- search journals by author, title, or keyword
- view article abstracts
- access and read full-text articles
- read forthcoming articles
- view and search references and related links
- download articles to a mobile device for offline reading
- capability to share links and article on social networks such as Facebook and Twitter

To access library subscriptions, users will need to visit http://annualreviews.org from their mobile device while they are authenticated via a wireless network. If connected to the subscribing library's wireless network, a user's device will automatically be paired with the library's access rights, which would enable them to access content from Annual Reviews while they are outside of the library's network. The pairing lasts six months and is renewed each time a user visits the Annual Reviews mobile website. If users are unable to pair their mobile device over a wireless network, they can log on to a network-connected computer and request a pairing code that will allow them to manually pair their device to access their library's content.

Taylor & Francis

Taylor & Francis is an international publishing company based in the United Kingdom that focuses on publishing books and academic journals. It is an additional library vendor that has chosen to provide mobile access through a mobile web interface. Taylor & Francis's mobile website will automatically detect what type of device a user is accessing its content from and redirect the user to the appropriate interface. As a result, users need only access Taylor & Francis Online from their library's electronic resources page or by visiting http://www.tandfonline.com. The mobile website supports most iOS, Android, and Blackberry devices, but if a user's device is not supported, he or she will be redirected to the full website. The features of the Taylor & Francis mobile website include

- an optimized interface for browsing, reading, and searching
- the ability to connect to social networks such as Facebook, Twitter, and LinkedIn
- access to a library's content by pairing a mobile device

- the ability to create a favorites list
- the ability to save full-text articles to a mobile device for offline reading

To access a library's subscription a user can pair his or her device. The device pairing lasts for 180 days. If a user is authenticated through the library's wireless network, his or her device will be paired automatically. Users can also manually pair their device in the same manner as they can for Annual Reviews. They can log on to a network-connected computer and request a pairing code from Taylor & Francis. There are detailed instructions on the Taylor & Francis website as well as a video demonstration on how to manually pair a device to access library subscribed content.

BioOne

BioOne is a nonprofit collaborative that was created to address inequities in science, technology, and mathematics publishing. More than 1,400 institutions subscribed to BioOne. The organization provides subscribing libraries with access to more than 150 journals. It is another library vendor that has also decided to provide mobile access to its resources through a mobile web interface, BioOne Mobile (Figure 6.2). The link to the mobile website is the same as it is for the full website (http://www.bioone.org). Users accessing BioOne from a mobile device will be automatically redirected to the mobile interface. Features of the mobile interface include

- mobile-optimized interface for browsing, reading, and searching full-text content
- one-click access to your favorite articles and journals
- filtering, searching, and sorting
- context-based searching for related research discovery

BioOne Mobile is currently accessible on iOS devices running iOS 6.1 and later, and Android devices running Android 2.3 and later.

Users can pair their device to BioOne Mobile through their library's subscription. If BioOne is accessed over the library's Wi-Fi network, then the device will be paired automatically, and a pairing code will not be needed. However, if users want to pair their device when they are not connected to the library's Wi-Fi network, they will have to manually pair their mobile device. To do that users will need to access their library's BioOne content from a computer, laptop, or tablet by logging in to their library's network and go to the following link: https://www.bioone.org/action/mobileDevicePairingLogin. Users will then be required to log in with their BioOne account or create a new account to obtain a pairing code. Once the pairing code is

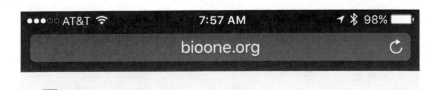

BioOne biology, ecology, and environmental science

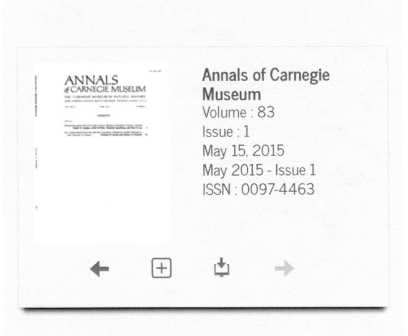

Annals of Carnegie Museum
Volume : 83
Issue : 1
May 15, 2015
May 2015 - Issue 1
ISSN : 0097-4463

Mammalian Petrosal from the Upper Jurassic Morrison Formation of Fruita, Colorado

New Rodent Material from the John Day Formation (Arikareean, Middle

Figure 6.2. BioOne Mobile

obtained, it will be valid for five minutes, during which the user will need to visit BioOne Mobile and tap on the Settings icon where they will select the Device Pairing option. The code will need to be enter and verified. Once the code is verified the mobile device will be paired with the account and library's subscription access rights for 90 days, after which the device will need to be paired again.

Web of Knowledge

Web of Knowledge is a research platform from Thomson Reuters that provides access to resources from the sciences, social sciences, arts, and humanities. It contains access to more than 23,000 journals. The platform includes the following resources:

- Arts and Humanities Citation Index
- Biological Abstracts
- BIOSIS Citation Index
- BIOSIS Previews
- Book Citation Index
- CAB Abstracts
- CAB Global Health
- Chinese Science Citation Database
- Conference Proceedings Citation Index
- Current Chemical Reactions
- Current Contents Connect
- Data Citation Index
- Dewent Innovations Index
- FSTA
- Index Chemicus
- Inspec
- MEDLINE
- Science Citation Index Expanded
- Web of Science
- Zoological Record

In order to provide libraries and their users with mobile access to these resources, Thomson Reuters has developed a mobile website (http://m.webofknowledge.com). In order to access the mobile website users must have a Web of Knowledge username and password. To create an account, users will need to log in to their library's network and access the Web of Knowledge platform and create a new account. This step is required to ensure that a user is from an authorized and subscribing library. Once the account has been created and verified, users will have access to the mobile version of Web of Knowledge.

Access to the mobile website can be achieved by typing in the mobile URL (http://m.webofknowledge.com) or by going to the full-site URL. If going to the full site, users will automatically be redirected to the mobile interface. The mobile version of Web of Knowledge has the ability to search within each of the individual resources included in the platform, as well as searching all the databases included. Other features of the mobile version include

- the ability to sort, refine, and e-mail
- the ability to add records to EndNote Web
- the ability to link to full text
- openURL
- the ability to view times cited counts
- access to citation score card
- access to search history

In order to keep access to the mobile version active, users are required to log in to Web of Knowledge from an IP-authenticated computer or device every six months to confirm that they are still an authorized user.

APPS

While some library vendors have chosen to provide mobile access through mobile web interfaces, other library vendors have chosen to provide mobile access to content for libraries and users through the development of mobile applications that target specific mobile platforms such as iOS and Android. Some vendors that have chosen this path are EBSCO, American Chemical Society, Mango Languages, Overdrive, LexisNexis, Gale, BrowZine, and Boopsie, among others.

EBSCOhost

Along with developing a mobile web interface for many of their databases, EBSCO has also developed several native applications that target both the iOS and Android mobile operating systems. One of those applications is the EBSCOhost app that is freely available to download in the App Store and Google Play. To download and authenticate the EBSCOhost application for iOS and Android, users need to navigate to an EBSCO database interface such as Academic Search Complete and click on the link at the bottom of the page that says "iPhone and Android Apps." A pop-up screen appears asking users for their e-mail address so EBSCO can send the authentication key for the application. Once an e-mail address is entered, users will get the following e-mail in their inbox with instructions on how to authenticate and access their library's collections.

Dear EBSCOhost user,
    To begin using the EBSCOhost iPhone and Android applications, follow the instructions below.

Step 1: Download the app from the iTunes Store or Google Play.
Step 2: View this e-mail on your device, then tap this authentication key.

(Note: You must access the link in Step 2 from your device. The activation link will expire in 24 hours.)
    Questions? Visit the support page or send an email to support@ebsco.com.

Thank you!
EBSCO Publishing

    Once authenticated, users will have access to the databases their library subscribes to through the EBSCOhost application. With the application users can

- choose which EBSCO database to search
- limit results to full text or peer reviewed
- get full-text results in HTML and PDF formats
- save results for offline access
- e-mail results
- automatically save the 25 most recent searches
- view overflow display of results

Updates to the application (version 3.0) include the ability to save PDFs to third-party applications such as Dropbox and iBooks and the option to e-mail saved articles along with some other minor enhancements.

EBSCO eBooks

In addition to the EBSCOhost application, EBSCO has developed and released (on August 11, 2015) and application for their EBSCO eBooks collections (Figure 6.3). Prior to the development and release of this

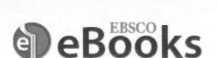

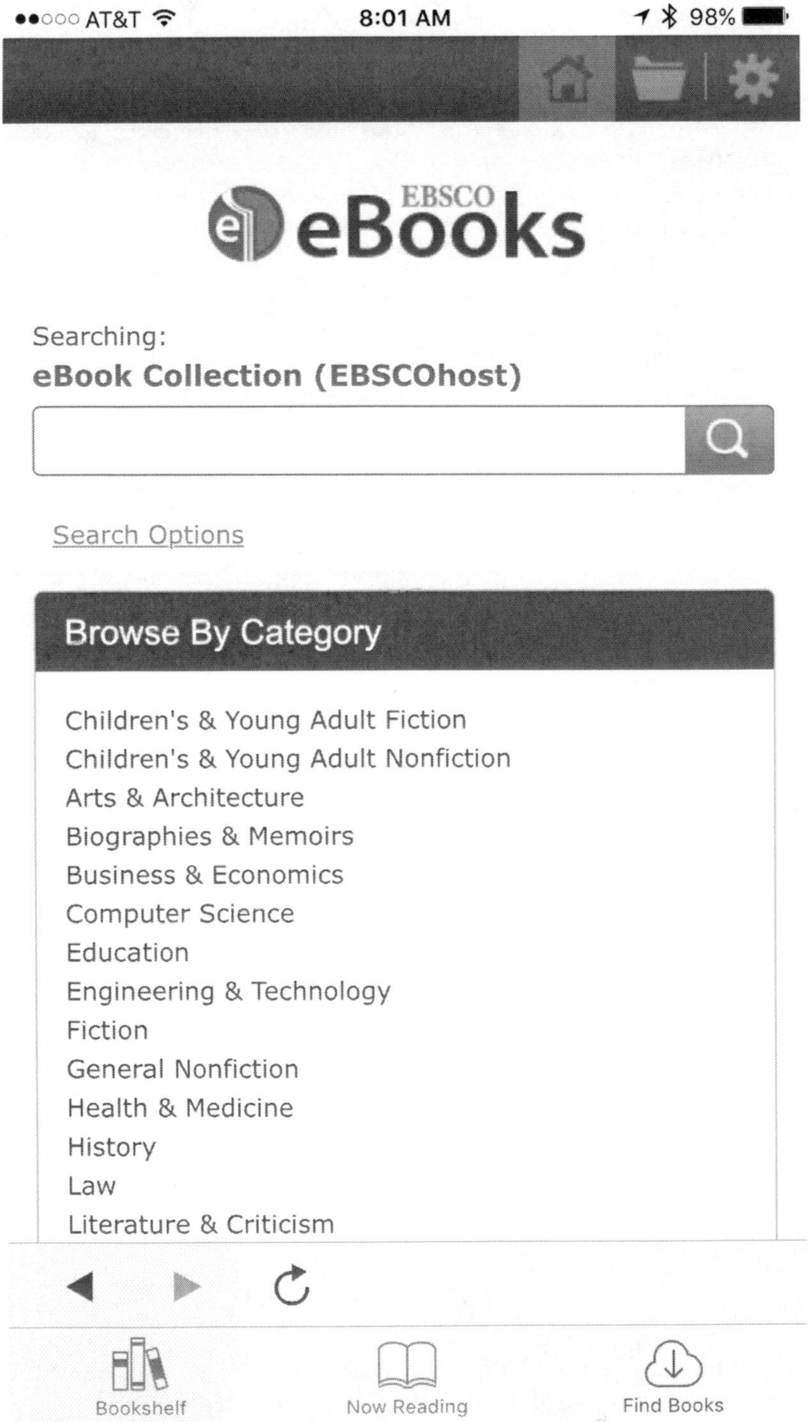
AT&T
8:01 AM
98%
EBSCO eBooks
Searching:
eBook Collection (EBSCOhost)
Search Options
Browse By Category
Children's & Young Adult Fiction
Children's & Young Adult Nonfiction
Arts & Architecture
Biographies & Memoirs
Business & Economics
Computer Science
Education
Engineering & Technology
Fiction
General Nonfiction
Health & Medicine
History
Law
Literature & Criticism
Bookshelf
Now Reading
Find Books
Figure 6.3. EBSCO eBooks App

application libraries and their users had to use a third-party application, like Bluefire Reader, to access EBSCO eBooks via a mobile device. The app is freely available on iOS and Android through the App Store and Google Play.

Once the application is downloaded users can search for their library and authenticate to access the library's e-books. With the EBSCO eBooks application, users can

- search and browse the library's e-book collection
- place holds on titles
- download e-books to the app and read offline
- customize their reading experience by adjusting font, brightness, and more
- return books

In order to use the application users will need to enter their Adobe ID or create one prior to authentication.

American Chemical Society

American Chemical Society (ACS) Publications is one of the most authoritative and comprehensive content providers for chemistry. In order to make its publications more accessible, ACS has developed a native application, ACS Mobile, which is freely available for the iOS and Android mobile operating systems. According to the descriptions in the App Store and Google Play, the features of the ACS Mobile application include

- Up-to-the minute access to new ACS ASAP Articles, personalized across the entire portfolio of over 40 peer-reviewed ACS journals.
- Tailored "on the fly" filtering options for viewing content from selected ACS titles.
- Delivery of an indexed list of more than 38,000 research articles published annually, complete with graphical and text abstracts.
- Automatic saving of abstracts for offline reading.
- A "Latest News" feed from *Chemical & Engineering News Online*.
- Saving of favorites in a "My ASAPs" folder for convenient offline reading and pushing back to your usual research setting.
- Interface to full-text article access (via wireless or virtual private networks) for users at institutions that subscribe to ACS journals. ID/password-based access is also an option for individuals who subscribe to ACS journals as part of their ACS member benefits package.
- Caching of full-text article PDFs for 48 hours to read offline.

- Quick search across the more than 850,000+ scientific research articles and book chapters now on the ACS Web Editions Platform—discoverable by author, keyword, title, abstract, DOI, or bibliographic citation.
- Sharing of links and snippets via e-mail, Facebook, Twitter and other options.

For users to authenticate and get full-text access to content that their library subscribes to, they need to be logged in to the wireless network or have an external login to access the library's system, such as a virtual private network login.

Mango Languages

Mango Library Edition is a subscription product from Mango Languages that includes a variety of resources to help library users to learn conversation language skills. Mango Mobile Library Edition is the application that the company has developed. It is available as a free download in the App Store and Google Play. The application works on iOS devices running iOS 4 or later and Android devices running Android 2.2 or later.

In the application iOS users will have access to the same resources their library subscribes to and that they are able to access in the online version through the library's website. Android users will have access to the same languages with the exception of languages that are written from right to left or languages with unsupported characters. Mango offers a variety of language courses for English speakers that include Arabic, Chinese, Croatian, Czech, Danish, French, German, Greek, Hebrew, Italian, Korean, Latin, Polish, Russian, and many more.

To use the application, users will need to create an account with Mango Languages via their library's website. Those account credentials will be used to log in to the application. Additionally, the application will be updated each time a user logs in if the person has Internet access.

OverDrive

OverDrive is a distributor of e-content such as e-books, audiobooks, music, and video. The content is delivered with digital rights management protection and allows library users to download available content. They work with a variety of different publishers, including Random House, HarperCollins, AudioGO, Harlequin, and Bloomsbury. OverDrive hosts more than 2 million digital titles from these publishers. OverDrive has also signed an agreement with Warner Brothers and other studios to build on their offerings of streaming video. In addition to providing this service to a large number of libraries, OverDrive offers an application on the Android and iOS platforms for users to access their library's OverDrive content. The application is available as a free download in both the App Store and Google Play.

The OverDrive application is compatible with iOS devices running iOS 6 or later and Android devices running Android 4.0 (also known as Ice Cream Sandwich). It is also available in the Amazon App Store, on Windows Phone, and as a Nook app. The features of the application include the ability to search the OverDrive content your specific library has access to. Titles that are downloaded in the application are automatically returned once the checkout time has expired. Users can create wish lists and place holds on items in the application. In order to use and access the content, users need a valid account with a participating library. To log in, users select their library and enter their library card number to be authenticated. Once authenticated, access to that library's OverDrive content will be accessible.

LexisNexis

LexisNexis has a wide variety of mobile solutions for their subscribers. They have applications for legal research, new and emerging issues, practice management and marketing, and litigation. Many of the LexisNexis applications have targeted the iOS platform with a few applications being available on the Android platform. One application that is available from LexisNexis is Lexis Advance HD. This application is designed specifically for the iPad and is only available on the iOS platform. There is also a version for the iPhone, which is Lexis Advance.

The Lexis Advance HD application is a tool that allows subscribers to access legal research from the Lexis Advance resource. Subscribers of this resource can

- view and annotate documents offline and sync changes
- view alerts
- access previous research results and search within those results
- organize with remote access to saved files and folders

Lexis Advance HD is available as a free download. Users must have a current subscription with a valid username and password to utilize the application.

This is just one example of an application that LexisNexis has developed. Other applications that they have developed include CourtLink (iOS and Android), LexisNexis Law School Q&A Series (iOS and Android), Lexis-Nexis Legal News (iOS), Nexis News Search (iOS), Lexis Legal News Briefs (iOS), Mealey's Legal News (iOS), LexisNexis Tax Law Community (iOS), LexisNexis Lead Alert (iOS and Android), Martindale-Hubbell Competitive Essentials (iOS), Martindale-Hubbell Lawyer Index (iOS), and LexisNexis TextMap App for iPad (iOS).

Gale

Gale is part of Cengage Learning, which provides a wide variety of teaching learning, and research solutions for libraries of all types. In addition to providing these solutions, Gale has developed applications for library users to access content from their subscribing library directly from their mobile device. One application that Gale has developed is AccessMyLibrary. There are different versions of this application depending on the type of library. There is the AccessMyLibrary School Edition, College Edition, Public Edition, and Special Edition.

The authentication process for each of the applications depends on the version a user is accessing. For example, the AccessMyLibrary Public Edition uses GPS positioning to determine the libraries that are within a 10-mile radius and then provides access to the user. The School and College Editions allow the user to select their school or college based on state location. Once a subscribing school is selected, the user will be prompted to enter his or her institution's password in order to gain access to the resources. The Special Edition operates the same way as the School and College Editions. Users select a participating library from a list and provide the specific password in order to gain access to the resources. Once authenticated through these applications, library users will have access to the Gale resources the library subscribes to. The versions of these applications are available as free downloads in the App Store and Google Play.

BrowZine

BrowZine is a free mobile application from Third Iron that allows users to browse, read, and monitor journals available through their library. This application (Figure 6.4) serves as an interface solution that allows users to access electronic journals in one place that their library subscribes to. The application is available in the App Store, Google Play, and the Kindle App Store. With the BrowZine application users can

- create a bookshelf of favorite journals for easy, fast access
- read articles in a format optimized for a mobile device
- get alerts when new journal issues are published
- save articles to Zotero, Dropbox, or an app of his or her choice for pdf collecting and notetaking
- share links to articles with others by e-mail, Facebook, or Twitter

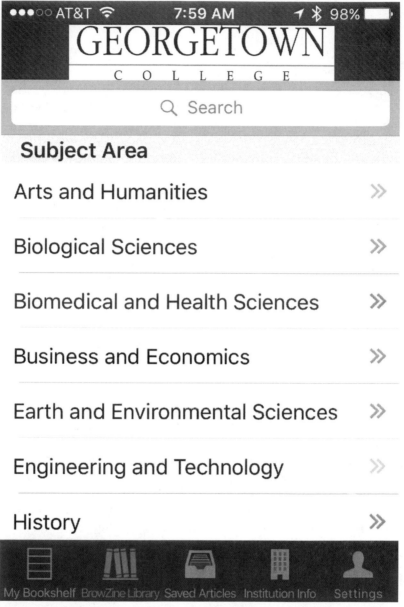

Figure 6.4. BrowZine

After the application is downloaded the user will be asked to "Choose a Library." Once the library is selected, the application will prompt users for their login credentials. Once they are authenticated, they can access the library's electronic content from supported publishers. As of version 1.6, BrowZine requires version iOS version 7 or newer to run on iPads, iPhones, and iPod

Touches. Android tablets and smartphones running version 4.1 (Jelly Bean) and the Kindle Fire HD are supported as well.

Boopsie

A library vendor that offers a service to develop branded native applications for libraries is Boopsie for Libraries. Boopsie was founded in Silicon Valley in 2006, and it has provided this service to more than 2,500 public and academic libraries, according to its website. Boopsie has also developed applications for library-related conferences such as the Library and Information Technology Association's National Forum. The applications that Boopsie creates for libraries are downloadable on a variety of different mobile devices and platforms, including Android, iOS, Windows Phone 8, Windows Mobile, Kindle Fire, Blackberry, Palm, J2ME, and Symbian Series 600.

Applications developed for libraries by Boopsie include features such as an integrated catalog search that allows users the ability to search the catalog from their mobile device. A library locator with GPS-aware technology is included that searches for branch locations, hours of operations, and contact information. The application integrates e-content delivery for e-books, e-audiobooks, and video content. Users are able to manage their accounts, which includes placing holds and renewing items, through the application with incorporation of the integrated library system. The Ask-A-Librarian feature is included which allows libraries to extend reference services to mobile devices with integration for text messaging, e-mail, phone, or library location information. Another feature included in a Boopsie-developed application is social media integration that allows users to stay current with the latest news and announcements from their library. These are just some of the features that are included in a Boopsie application. In addition to providing these features, Boopsie also provides a website where libraries and their users can demo the application before it is downloaded on their device.

Boopsie provides the option to deploy native applications to libraries that do not have the staff to develop applications within the library. Boopsie also stays current on mobile technology, which makes the maintenance of the application easier, and the application integrates with a variety of different integrated library systems. The service provided by Boopsie has been used by a variety of different libraries such as the Seattle Public Library, Centre College (KY), San Diego Public Library, and the Charlotte Mecklenburg Library. Boopsie is a library vendor whose focus is to provide mobile solutions for libraries of all types and was acquired by Demco, Inc., in 2015.

These are just a few examples of library vendors that have provided mobile access to their content for libraries and their users. Some have chosen to provide access through mobile web interfaces and others have chosen to develop native applications for specific mobile platforms. Regardless of the approach, library vendors have recognized the importance of mobile technology and the need and desire of users to access and consume content from the device of their choice. By providing these mobile solutions vendors are enabling libraries and users to choose their preferred format.

Chapter Seven

Wearable Technology

A rising category within mobile technology that has begun to gain popularity and traction over the last several years is wearable technology. While it is true that wearable technology has existed for some time, the new focus on wearable technology goes beyond the realm of what we have previously expected. Technology companies like Apple, Samsung, and Google are focusing efforts on developing devices and software packed with rich features that extend the use of current mobile devices or act as standalone pieces of wearable technology with their own unique features. With the massive success that mobile technology has brought, these companies are now looking for additional ways to expand the use of mobile technology beyond the realm of smartphones and tablets to make mobile technology something that is more intimate and, in some cases, fashionable. With this we have seen the emergence of new companies and products that have been rather successful, such as Fitbit and GoPro cameras. There are even some instances of companies looking at ways to infuse technology into clothing. Mobile technology has had a tremendous impact on the way that we communicate and interact with the world around us. The emergence of wearable technology provides some interesting ideas and potential as to how mobile technology will continue to influence our communication and interactions. With the emergence of wearable technology, it is important for libraries to start thinking about ways that these devices can be used to extend services and resources to patrons.

The emergence and potential of wearable technology has caused EDU-CAUSE to include it in their EDUCAUSE Learning Initiative "7 Things You Should Know About" series. The series provides concise information of emerging technologies and their implications on teaching and learning in higher education. Each part in the series focuses on a particular emerging technology. The publication addresses seven different questions that people need to be aware of.

1. What is it?
2. How does it work?
3. Who's doing it?
4. Why is it significant?
5. What are the downsides?
6. Where is it going?
7. What are the implications for teaching and learning?[1]

Before you can understand the implications and possibilities of a technology, you first need to know what it is. So, what is wearable technology? Wearable technology refers to devices that can be worn by users and can take the form of an accessory, such as jewelry or sunglasses, or physical items of clothing, such as shoes or a jacket. The benefit of wearable technology is that it can integrate tools, devices, power needs, and connectivity within a user's everyday life and movements.[2] This technology enables users to obtain a wealth of information regarding their surroundings.

How does wearable technology work? Generally, most wearable technology focuses on a narrow range of functionality with a limited set of features. However, there are examples of wearable technology that are more complex, multifunction systems, such as Google Glass.[3] Some of the functions of wearable technology include providing the user with notifications from text message and social media. The technology can also track fitness and health statistics such as steps taken during the day, calories burned, and heart rate. Some forms of wearable technology, like the Samsung Galaxy Gear smartwatch, also enable users to answer phone calls.

While wearable technology is still relatively new, there are some organizations that are experimenting with the potential uses. For example, the School of Medicine at the University of California, Irvine, has issued Google Glass to all of its medical students. UC Irvine is the first school to implement the use of wearable technology into its four-year curriculum. Dr. Ralph V. Clayman, UC Irvine's dean of medicine, said,

> I believe digital technology will let us bring a more impactful and relevant clinical learning experience to our students. Enabling our students to become adept at a variety of digital technologies fits perfectly into the ongoing evolution of health care into a more personalized, participatory, home-based, and digitally driven endeavor.[4]

However, many of the uses of wearable technology still remain experimental as institutions and organizations are still weighing the potential uses as well as the drawbacks.

The significance of wearable technology is that it represents an evolution in people's relationship with computing and hints at a future of continuous connectivity. Wearable technology presents a list of potential benefits that includes monitoring the health of the users, which could enable timely medical intervention.[5] This technology can also provide video records of events, and in some forms, it can incorporate an augmented reality overlay to real-life situations to enhance learning.

Regardless of the potential of wearable technology, there are some downsides. The most notable drawback is that the connectivity of wearable technologies, like Google Glass, raise questions of privacy, security, and information consent.[6] The main concern over privacy is that these devices can be hacked like any other device that can send and receive data. Other people can be recorded without their consent, and some of these devices, such as Google Glass, have been banned from certain establishments. In regard to the classroom, there are concerns about cheating and unauthorized recording of lectures. Additional concerns are that the power needed to run these new devices will mean heavier batteries and increased heat, along with the perception that some of these devices are not fashionable. Despite these drawbacks many experts predict that wearable technology will soon enter the mainstream; we are starting to see this with devices such as Fitbit, Samsung Galaxy Gear, and Apple Watch. In fact, wearable technology has become a part of the Mobile World Congress. The Mobile World Congress is an annual international event held in Barcelona that focuses on the new and emerging developments in mobile technology and their implications.

There are a variety of different ways that wearable technology could be going, depending on the aim of the device. EDUCAUSE Learning Initiative stated that wearable technology could incorporate haptic feedback, such as alerts to messages or upcoming appointments. The devices might also combine touch with location-based notification. Other devices might include biometrics for more effective security. An example is the Nymi wristband by Bionym. This wristband can unlock a laptop, smartphone, tablet, or bank account, and confirms the identity of the user by his or her unique heartbeat. More and more companies are developing wearable technology. Samsung has released different wearable options like the Galaxy Gear and the Galaxy Gear Fit. Apple has also entered into the wearable market with the development of the Apple Watch. These developments show that technology companies like Apple, Google, and Samsung see wearable technology as a category with great growth potential.

There is a wide array of implications for teaching and learning in regard to wearable technology. It has the potential to alter the landscape of educational computing by enabling the learner to engage in a variety of different functions. The EDUCAUSE Learning Initiative points out that wearable cameras would allow learners to engage as an observer, reporter, and participant. There is the potential for data gathering in biometrics and environmental conditions with less human interaction, which means less risk for contamination. Additionally, wearable technology devices could offer great assistance to people with visual, auditory, or physical disabilities.[7] These are far-reaching implications that have an effect on not only teaching and learning, but also the broader society. This in turn means that libraries will need to be aware of and start evaluating the impact wearable technology will have on their organizations. Some libraries are starting to experiment with forms of wearable technology such as Google Glass.

GOOGLE GLASS AND LIBRARIES

Google Glass, to this point, is probably the most notable of wearable technology. It is a wearable computer with an optical head-mounted display that is activated by voice commands. The device resembles a pair of glasses, but it has a single lens. It allows users to consume and engage with information about the surroundings, as well as collect data about their surroundings. Most of the efforts related to wearable technology in libraries centers on Google Glass. Libraries are starting to explore the potential impact that Google Glass may have on the research and information behaviors of their users.

Google Glass is not yet widely available to the general public. For the initial release, Google had an Explorer Edition contest, where users would post on Google+ or Twitter, in 50 words or less, what they would do if they had Google Glass. Several libraries were able to be one of the Google Glass "Explorers." One was Claremont Colleges Library. The library staff began exploring the teaching, learning, and research potential of Google Glass in the spring of 2014. During that time, they invited their users to submit proposals on how Google Glass might be used in teaching, learning, and research.[8] They also held hands-on workshops to allow their users to interact with the technology.

North Carolina State University Libraries, Yale University Library, and Arapahoe Library District in Colorado are additional examples of libraries who have access to Google Glass. North Carolina State University Libraries launched a pilot program in 2014 for selected faculty and graduate students who had an urgent research need for the technology. Requests were accepted from researchers working with augmented reality, innovative computer interfaces, and other possible uses.[9] Yale University Library took much of the same route as Claremont Colleges Library and North Carolina State University Libraries. They encouraged faculty and student groups to submit requests and proposals on the potential of Google Glass in enhancing classroom instruction and the research experience.[10] The Arapahoe Library District enables users to interact with Google Glass at their Goggle at Google Glass events. Adults, teens, and children six years and older are able to interact with the technology at the events. The Arapahoe Library District feels that Google Glass is an important technology to have in libraries. Their website states that "Google Glass represents the future of technology, and as a library we serve as an unbiased technological-literacy resource for patrons. We anticipate that Google Glass represents the first iteration of the 'next big thing' in gadgets, and we want our patrons to have the opportunity to try it for themselves."[11] In addition to allowing users the opportunity to interact with the technology, there are a variety of other potential uses of Google Glass in libraries.

An article by Ellyssa Kroski on the Open Education Database website listed seven different ways that libraries can use Google Glass.

1. Enhance libraries tours by adding augmented reality overlays and imagery, along with audio and video files that explain the history of the building, library collections, and more.
   Record author talks and library events.
2. Enhance Makerspaces by providing builders with the additional technology that could provide them with helpful information like diagrams and instructional videos.
3. Record hands-on video tutorials that give users the perspective of the creator's point of view.
4. Provide real-time optical character recognition (OCR) and text-to-speech translation for the visually impaired.
5. Provide real-time language translation of foreign texts.
6. Speak to users in their own language with the capability of voice translations.[12]

These are just a few examples of how Google Glass can be used in libraries. However, Google stopped selling Google Glass in January 2015. There is good news. Google has filed a report with the Federal Communications Commission for a new version of Google Glass whose design is similar to the older version, but the new device appears to be more suited as a business tool than a device meant for popular everyday use.[13] According to an article in the *Wall Street Journal*, the new version of Google Glass is being pitched as a device that is valuable to fields such as health care, manufacturing, and energy.[14] It is unclear what the implications for libraries might be, but the possibilities provided by a device like Google Glass are intriguing, especially since some libraries have already had the opportunity to implement it.

OCULUS RIFT

Oculus is a virtual reality technology company that makes the Oculus Rift headset. It was founded by Palmer Lucky through a Kickstarter campaign that raised $2.5 million in 2012. Less than two years later, Facebook bought the company for $2 billion. The deal included $400 million in cash and 23.1 million shares of Facebook common stock (based on the stock price of $69.35). The deal also allowed for an additional $300 million in earn-out cash and stock based on milestones. Facebook CEO Mark Zuckerberg explained the reason for the acquisition: "Mobile is the platform of today, and now we're also getting ready for the platforms of tomorrow. . . . Oculus has the chance to create the most social platform ever, and change the way we work, play, and communicate."[15] For the best experience using Oculus Rift, Oculus recommends the following PC system:

- Graphics card: NVIDIA GTX 970 / AMD R9 290 equivalent or greater
- Processor: Intel i5–4590 equivalent or greater
- Memory: 8GB+ RAM
- Output: Compatible HDMI 1.3 video output
- Input: 3x USB 3.0 ports plus 1x USB 2.0 port
- Operating system: Windows 7 SP1 64 bit or newer

Oculus also plans to start making bundles with Oculus Ready PCs and Rift available in February 2016. In addition to the Oculus Rift headset, Oculus also develops the Gear VR headset for Samsung. The Gear VR can be used with certain Samsung mobile devices such as the Galaxy Note 5, Galaxy S6, Galaxy S6 Edge, and Galaxy S6 Edge+.

Some libraries have started to acquire these headsets to make them available for their users to test out. Arapahoe Library District in Englewood, Colorado, is one library that has acquired Oculus Rift. On its website, the library addresses why it invested in technologies such as Oculus Rift:

> As a resource for the community, the library offers cutting-edge, new technology that may be slightly out of reach as a way to give our patrons a chance to try out technology before buying it, and remain informed. We are also better able to stay up on technology that way, and assist the public in a non-biased way. Unlike a retail store, we have nothing to sell, so we can offer advice and support on a variety of devices. We look at this as an extension of community literacy—technological literacy, which is extremely important in the world we live in.[16]

The Arapahoe Library District has at least one Oculus Rift at its eight branches. Users wanting to test out the device can do so by "booking a librar-

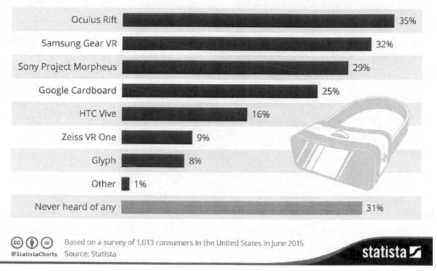

Which VR Headset Holds the Pole Position?

% of U.S. consumers (ages 19-49) who have heard of the following VR headsets

Headset	%
Oculus Rift	35%
Samsung Gear VR	32%
Sony Project Morpheus	29%
Google Cardboard	25%
HTC Vive	16%
Zeiss VR One	9%
Glyph	8%
Other	1%
Never heard of any	31%

Based on a survey of 1,013 consumers in the United States in June 2015
@StatistaCharts   Source: Statista

statista

Figure 7.1.   Virtual Reality Headsets. *Statista.com*

ian." It also has a web page devoted to answering questions users may have about Oculus Rift. Other libraries that have the Oculus virtual reality headset available are North Carolina State University Libraries, University of Wisconsin–Madison Libraries, Georgia Institute of Technology Libraries, and Arlington Heights Memorial Library in Illinois.

There are a few other companies that are working on virtual reality headsets to compete with Oculus. Most notably at the moment is Sony with its virtual reality headset, Project Morpheus. Apple has filed for a patent for a head-mounted display, but the company does not seem to have any plans to release a virtual reality headset in the near future. Oculus Rift is currently the most recognizable virtual reality headset (Figure 7.1). Virtual reality offers a host of new opportunities for communication and interaction. As this type of wearable technology continues to develop, it will be interesting to see the different applications for use in libraries.

GoPro

GoPro cameras are another wearable device that are starting to make their way into libraries. These devices are micro-cameras that have the ability to record video and capture still photography, and they have burst capabilities. GoPros are designed to endure rough conditions while the wearer takes action shots and video. We often see these cameras being worn, usually on a helmet, by people riding in extreme sporting events such as motocross or driving an ATV through rugged terrain or through a desert. The current flagship camera from GoPro is the Hero 4. Features of the GoPro Hero 4 include

- 2× more powerful processor compared to the Hero 3+
- Ultra high-resolution featuring 4K30, 2.7K60, 1080p120, and 720p240
- 12MP camera that allows for capturing photos at 30 frames per second
- New Night Photo and Night Lapse modes
- Waterproof to 131 feet

These cameras can be worn with various mounts for the head and chest. GoPros offer a head strap and clip as well as a chest harness for users to wear the device to capture videos and images. With built-in Wi-Fi and Bluetooth you can control, view, and share content created with GoPro cameras from the GoPro App and Smart Remote.

Libraries that are making these devices available for users include Sacramento Public Library, University of Miami Libraries, Colorado State University Libraries, and the University of Memphis Libraries. The University of Memphis has a LibGuide about the GoPro cameras available at its library. The university allows users to check out the devices for three days with no renewals. Late fees are $5 a day. If users are not familiar with a GoPro, they can attend a workshop. The LibGuide also covers how to use the GoPro app and what mounts are available for checkout.

Berea College's Hutchins Library has looked to incorporate a GoPro by experimenting with creating a library tour using the device. Amanda Peach, the instructional service librarian at Berea, says that there were a few problems using the GoPro for creating a library tour. One was that the GoPro had trouble picking up the voices of students as they were trying to act out a skit. The device was able to pick up the voice of the student wearing the GoPro but not the others. Another issue was that as the students talked (and were still), the camera would zoom in and out trying to focus on various items or people as they passed. Amanda mentioned that she felt these issues could be solved by having students go around the library and take actions shots of people doing things, such as students laughing, students studying, instructors teaching, and someone shelving a book. She stated that the GoPro is best suited for capturing action. She is planning to experiment more with creating a library tour using the GoPro by taking action shots and looking to add a voice-over dialogue at a later point.[17] These are just some examples of how libraries are using GoPro cameras by making them available to patrons and looking at ways to use them to create interactive videos.

SMARTWATCHES

The most promising category in wearable technology according to many forecasts and predictions revolves around the wrist. Wearable technology that focuses on the wrist is considered to be the future of the category (Figure 7.2). While many of the earlier Smartwatches focused on fitness tracking, such as the Fitbit, companies like Samsung and Apple are looking at ways to expand the use of smartwatches. What has come from those efforts are the Samsung Galaxy Gears (the Gear S2 being the most recent) and the Apple Watch. Currently these devices are extensions of compatible smartphones. They enable users to receive notifications such as text messages and calendar events. Users can also receive and make phone calls with these devices as well as track fitness activities. The specifications for the Samsung Galaxy Gear S2 include

- A 1.2-inch full circular sAMOLED display at 302ppi
- 512MB of RAM and 4GB of storage
- Bluetooth and 3G network
- Bluetooth 4.1, Wi-Fi, and NFC (near field communication) connectivity
- Built on Samsung's Tizen Operating System

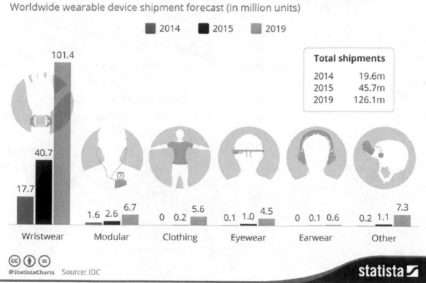

Figure 7.2. **Predicted Wearable Technology Boom Is All about the Wrist.** *IDC / Statista.com*

The Apple Watch comes in three different collections: Watch Sport, Watch, and Watch Edition. The specifications for the Apple Watch include

- A 38mm (340x272 pixels) display and a 42mm (390x312 pixels) display
- 8GB total storage with 2GB for music and 75MB for photos
- Bluetooth 4.0 Low Energy and Wi-Fi connectivity
- Built on Apple's new WatchOS

According to numbers from *Forbes* in November 2015, Apple sold 7 million Apple Watches since it launched in April 2015, although Apple has launched an official numbers regarding sales of the device. *Forbes* also stated that Samsung smartwatches managed to sell 300,000 units in the same quarter.[18] Competitors to Apple and Samsung continue to look for ways to become a major player in the smartwatch category. Fitbit is one example. The company recently released a smartwatch, the Fitbit Blaze, that offers some of the same functionality as the Apple Watch and Samsung Galaxy Gears to extend its reach and perception beyond that of a wearable device that just tracks fitness activity. Other wearable devices include the Pebble Smartwatch and the Microsoft Band.

The Consumer Technology Association predicts that 38 million wearable devices will be sold in 2016 with activity trackers and smartwatches making up a majority of those sales. However, it appears we are still at a point where smartwatches are not seen as universally useful to people, but are viewed, rightfully so at this point, as a second screen for a smartphone. Dan Ledger, principal for Edeavour Partners, said that smartwatches are "a fun thing to have . . . but it's not nearly as essential of a device as smartphones and for some people an activity tracker." Speaking along the same lines, J. P. Gownder, vice president and principal analyst at Forrester, stated that "smartwatches will continue to grow, bit they'll also continue to disappoint some people. They're never destined to become as common as smartphones. . . . Instead, they'll become a common smartphone accessory that reaches 25 million people globally by the end of 2016."[19] While they may not become as essential or common as a smartphone, smartwatches do offer a lot of advantages to users. As tech companies continue to put more effort and resources into the development of these devices and external developers become more aware of their capabilities, these devices will become more useful.

Smartwatches like the Apple Watch and Samsung Galaxy Gear 2 offer some interesting possibilities for libraries. These devices offer a way for libraries to extend notifications to users. Possible ways to do that may include notifying users about events when they enter the library, when they have a book on hold that is ready for pickup, and when they have an item that is due; it could also include helping users navigate the stacks to find a book. With voice command and text message capabilities, users could even use these devices to contact library staff with questions about resources and services. Some libraries are currently experimenting with checking smartwatches out to users, such as North Carolina State University Libraries, which has the Apple Watch available for checkout. As these devices become more popular and mainstream, it will be interesting to see the different ways that libraries leverage these mobile devices to extend resources and services to users.

CONCLUSION

Wearable technology provides a lot of potential for libraries. People addressed the question of how wearable technology might be relevant to academic and research libraries on the "Horizon Report: 2014 Library Edition Wiki"; responses included

- Wearable technology could facilitate energy efficient heating/cooling/lighting/security in library spaces by registering the number of people using parts of the building and adjusting energy accordingly.
- Language translations—for communicating with patrons and also reading foreign language books, websites, etc., accessibility for sight or hearing impaired for example books read aloud, etc., new augmented reality apps will make these devices even more powerful by providing added layers of information over the real world viewed through devices such as Google Glass, etc. These will be powerful tools for librarians as well as to offer patrons through lending programs and/or non-circulating use once the price point comes down.[20]

Responses also included statements pertaining to privacy concerns, and that libraries and librarians need to be aware of these issues since some forms of wearable technology are essentially cameras. Libraries also need to be aware that many of these devices collect data and information about the surroundings of the user. Despite this, one responder noted that wearable technology is the direction that things are going in. As a result, libraries will need to be knowledgeable even if they are not on board with these devices. Regardless of where this technology goes, it offers another medium for libraries to reach users. These new devices do bring many challenges, but they also offer tremendous possibilities.

NOTES

1. EDUCAUSE. "7 Things You Should Know about . . . Wearable Technology." EDUCAUSE Learning Initiative, November 2013. https://net.educause.edu/ir/library/pdf/ELI7102.pdf

2. New Media Consortium. "NMC Horizon Report: 2014 Library Edition Wiki." 2014. http://lib-2014.wiki.nmc.org/Wearable+Technology

3. EDUCAUSE. "7 Things You Should Know about . . . Wearable Technology."

4. Kerr, D. "Google Glass Handed Out to Medical Students at UC Irvine." CNET, May 14, 2014. http://www.cnet.com/news/google-glass-handed-out-to-all-medical-students-at-uc-irvine/?ttag=fbwl

5. EDUCAUSE. "7 Things You Should Know about . . . Wearable Technology."

6. Ibid.

7. Ibid.

8. Claremont Colleges Library. "Google Glass @ the Claremont Colleges Library." Claremont University Consortium, January 12, 2014. http://libraries.claremont.edu/glass.asp

9. North Carolina State University Libraries. "NCSU Libraries Now Lending Google Glass for Research Projects." NSCU Library News, February 20, 2014. http://news.lib.ncsu.edu/2014/02/20/ncsu-libraries-now-lending-google-glass-for-rcsearch-projects/

10. Yale University Library. "Google Glass available for faculty and student groups during spring semester." January 31, 2014. http://www.library.yale.edu/librarynews/2014/01/google_glass_available_for_fac.html

11. Arapahoe Library District. "Goggle at Google Glass." Arapahoe Libraries, 2014. http://arapahoelibraries.org/googleglass

12. Kroski, E. "7 Things Libraries Can Do with Google Glass." University of Cyprus Scripta Pteroenta Blog, April 22, 2013. http://libblog.ucy.ac.cy/2013/04/7-things-libraries-can-do-with-google.html?m=1

13. Dolcourt, Jessica. "Google Glass 2.0 Is Real, and Here Are Photos to Prove It." CNET, December 28, 2015. http://www.cnet.com/news/google-glass-2-0-is-real-photos/

14. Barr, Alistair. "Google Quietly Distributes New Version of Glass Aimed at Workplaces." *Wall Street Journal*, July 30, 2015. http://www.wsj.com/articles/google-quietly-distributes-new-version-of-glass-aimed-at-workplaces-1438283319

15. Van Grove, Jennifer. "Facebook to Buy Oculus for $2 billion." CNET, March 25, 2014. http://www.cnet.com/news/facebook-to-buy-oculus-for-2-billion/

16. Arapahoe Library District. "Oculus Rift." Arapahoe Libraries, 2015. http://arapahoelibraries.org/oculus-rift

17. E-mail correspondence with Amanda Peach (Berea College's Hutchins Library).

18. Lamkin, Paul. "Apple Watch Sales Hit 7 Million." *Forbes*, November 5, 2015. http://www.forbes.com/sites/paullamkin/2015/11/05/apple-watch-sales-hit-7-million/#23c2ec225c9a55dca6f75c9a

19. Maddox, Teena. "Top IoT and Wearable Tech Trends for 2016: Smartwatches in Transition as Smartglasses Rule." January 14, 2016. http://www.techrepublic.com/article/top-iot-and-wearable-tech-trends-for-2016-smartwatches-in-transition-as-smartglasses-rule/

20. New Media Consortium. "NMC Horizon Report: 2014 Library Edition Wiki." 2014. http://lib-2014.wiki.nmc.org/Wearable+Technology

Chapter Eight

The Future of Mobile Technology

The rise and influence of mobile technology over the last decade is, arguably, unlike anything we have seen with other forms of technology. It has changed the way that we consume information and how we interact with others and the world around us. It has provided another platform for businesses, organizations, and libraries to reach consumers and users. For many mobile technologies, mobile devices in particular have become the first and main source of information access. With the continued innovation in mobile technology and the introduction of new device categories, libraries face both a challenge and an opportunity. The future of mobile technology is bright, and libraries must be willing to embrace this to remain relevant to our users. Some libraries are taking advantage of mobile technologies that will shape the future of mobile.

NEAR FIELD COMMUNICATION

A feature that is becoming more prevalent in mobile devices, especially smartphones, is near field communication (NFC). NFC is a set of short-range wireless technologies, typically requiring a distance of four centimeters or less to initiate a connection. It allows users to share small bits of data between an NFC tag (similar to QR codes, but with more storage capacity) and a device with an NFC chip, or between two devices with NFC chips. The technology is currently available in different Android-powered devices and Apple devices like the iPhone and the Apple Watch, but Apple devices currently only use NFC technology as part of the Apple Pay system. Android powered device with NFC support three main modes of operation.

1. Reader/writer mode, allowing the NFC device to read and/or write passive NFC tags and stickers.
2. P2P mode, allowing the NFC device to exchange data with other NFC peers; this operation mode is used by Android Beam.
3. Card emulation mode, allowing the NFC device itself to act as an NFC card. The emulated NFC card can then be accessed by an external NFC reader, such as an NFC point-of-sale terminal.[1]

NFC offers a lot of opportunities for businesses as well as individual users. This technology can enable people to pay for purchases using their mobile devices, and they can share photos with their friends by touching the devices together. There have been various experimentations with NFC at some higher education institutions. For example, Arizona State University partnered with HID Global to create a pilot on the use of NFC-enabled smartphones as mobile keys for building access.[2]

In addition to the applications and implications for businesses, libraries have also considered the possible uses of NFC. Some libraries have used NFC tags to enable NFC-capable devices to access information. NFC tags are similar to QR codes, but they are able to hold more information. The difference is that many mobile devices can download a QR code reader and scan a QR code, while NFC tags require a device that has the technology already built in. This in turn means that it is less accessible, since a majority of mobile devices still are not equipped with NFC. However, there are still some practical applications of NFC for libraries.

One possible application is that mobile tagging allows for easier searching, locating, and renting of books. Information embedded could contain bibliographic information, links to similar resources, and the due date of a book that is checked out.[3] The Hanno library in Japan is an example of a library that has used NFC in this capacity. Their library installed around 100 NFC tags that direct users to Wikipedia links to authors, pictures, and reviews.[4] Additional uses of NFC tags include faster access to e-books, and the replacement of traditional keys and library cards. These are a few examples of how libraries can use NFC technology to interact with mobile users. As more mobile devices become equipped with NFC technology, libraries will look at new and innovative ways that this technology can be infused into library services.

iBEACON

Another mobile technology that is starting to make waves is iBeacon. Trademarked by Apple, iBeacon is a technology that extends location services in iOS. It is an indoor positioning system that enables iOS devices running iOS 7 and later, as well as other hardware, to send push notifications to other iOS devices in close proximity. Android devices are able to receive these notifications, but are unable to send notifications via iBeacon. This technology enables retailers to send notifications, such as specials and coupons, to customers who are in close proximity to their establishments. The technology is still relatively new, but is being tested in a few different capacities. Major League Baseball (MLB) is implementing the technology, and 20 of the 30 MLB teams are participating. Fans that use the MLB at the Ballpark app will be able to receive various notifications at different points in the stadium. The notifications can help fans find their seats, receive discounts on merchandise and concessions, and queue up videos.[5] The San Francisco Giants were the first team to implement the technology. Bill Schlough, their chief information officer, stated, "It's kind of a no-brainer," when he was asked why they decided to implement it. He also stated that "mobile and digital experiences are paramount to our fan experience . . . and they have played a role in the fact that we've had 246 straight sellouts."[6] Retailers like Macy's have also used this technology. Libraries are starting to venture into the use of beacon technology as a way to reach users with mobile devices. Two vendors that offer libraries iBeacon based solutions are Capira Technologies and BluuBeam.

Capira Technologies

Capira Technologies is a company that develops mobile applications for libraries and is starting to integrate iBeacon technology in their applications to extend the reach of libraries to mobile users. Some of the features that iBeacon makes possible are

- Circulation notices—Patrons who have authenticated their account information in your library app can receive notifications about items due that day, items ready for pickup, and much more when they enter the building.
- Event notices—A patron can walk into your library's teen section or a specific branch building and get information about events happening that day for that location. Libraries can customize these notices; it could display a list of upcoming events for that section, or just remind patrons of events upcoming in the next hour or two.
- Informational notices—Patron devices can launch an informational notice about a specific area in the library and items found within a certain area if their device finds the beacon located nearby.
- Shelving notices—A patron can put his device near a particular shelf and see a list of items located in that section. For example, if a library offered a

row of shelves with New Releases, a patron could view items released that day using their device and a beacon located on the shelf.
*   Patron assistance—Devices can time how long a beacon stays in range. Staff can be notified if a patron spends an excessive amount of time in a specific area or room without moving, possibly indicating they may require assistance looking for items.
*   Beacon tracking—Anonymous beacon tracking allows libraries to capture how patrons with the library app move throughout the building, along with time spent in each area.[7]

The incorporation with the integrated library system and event calendar platforms is what makes some of these features possible.

Capira Technologies has also done some beta testing on a beacon feature that will send general notifications to users. Michael Berse of Capira Technologies stated that "if someone walks in to the library, it will check whether [he or she has] the app on [his or her] phone, then it will go out and check the patron record for certain things and say, 'Hello, you have two items due today. Visit the circ desk to renew them.' Or, 'You have two items ready for pickup.'"[8] Libraries that are using CapiraMobile are Somerset County Library System in New Jersey and the Half Hollow Hills Community Library in New York, both libraries were beta testers for the iBeacon integration.

BluuBeam

BluuBeam enables libraries to interact with users through the use of beacon technology by providing both hardware and software. The company was founded out of a desire to provide public libraries with a better way to interact with users. The hardware that BluuBeam provides are small circular discs that use beacon technology to enable libraries to interact with users based upon their location in the library. The software allows libraries to schedule different announcements that the beacons relay to users. With BluuBeam, libraries can make users aware of different resources and services available as well as what type of events are going on in the library. Users are made aware of this through the BluuBeam app that is available on Android and iOS.

The Orange County Public Library in Orlando, Florida, has implemented BluuBeam. It has installed beacons that transmit information both to users inside the library and to those walking past the building. This gives the library the opportunity to reach users who may not typically come to the library by strategically placing the discs in locations to reach users. This could include alerting people that are walking by the building of upcoming events scheduled in the library. A beacon could transmit to the user that the library has a streaming video collection. BluuBeam enables library users to save information that they are interested in or forward a beam onto a friend.

Fayetteville Free Library (FFL) in New York is another library that has implemented BluuBeam. They are using it to duplicate event notifications and have positioned one beacon so that it sends notifications to people outside of the library. Executive Director Sue Considine said that BluuBeam is a good fit for their library: "What we experience frequently here at FFL—and why this is such a great fit for us—is that we have high volume, high quality programming all the time, yet we get constant feedback from patrons saying, 'I didn't know about that.' Or, 'I didn't hear about that, despite the library's outreach efforts via newsletters, social media, its website, and other means.'"[9] This can help eliminate that issue.

Both CapiraMobile and BluuBeam need users to opt-in to the service, and both can be set up to send each notification to a user to a unique device only once. This ensures that users will not continue to be sent multiple notifications in the same location. Google, like Apple, has developed its own beacon technology and released the code as open source on GitHub called Eddystone.

EDDYSTONE

Eddystone is an open protocol released by Google that defines a Bluetooth low-energy message format for proximity beacon messages. The design of Eddystone was driven by key goals that include

- working well with Android and iOS Bluetooth developer APIs
- straightforward implementation on a wide range of Bluetooth low energy devices
- flexible architecture permitting development of new frame types
- full compliance with the Bluetooth Core Specification[10]

While the goal of both Eddystone and iBeacon are similar, there are some significant differences. Much like the difference between Android and iOS, Eddystone is an open-source protocol with the code available on GitHub under the Apache v2.0 license, while iBeacon is a closed technology controlled by Apple and only available on iOS devices. Eddystone has official support for both Android and iOS. Another differentiating feature of Eddystone is that it supports multiple frame types, or data payloads, while previous beacon solutions have served only one purpose. These frame types include a Universally Unique Identifier (UUID), URLs, Ephemeral Identifiers, and telemetry data.

The UUID identifies each specific beacon that an app can recognize and perform certain actions for. It can help organizations send messages to users in range of a certain beacon. The UUID is what Apple's iBeacon sends out. Sending a URL through is more universal and easier. The UUIDs are considered the QR code version of a beacon. Ephemeral Identifiers are more secure. They are similar to a personal beacon that only authorized users can read. Telemetry data are intended for organizations that need to manage a large amount of beacons. This frame type would send diagnostic data and battery life to an IT department so that it could manage and fix the beacons if needed.[11]

Libraries are starting to take advantage of beacon technology to reach users through their mobile devices both inside and outside the library. As the technology continues to develop and improve, it will be interesting to see the impact that it will have on libraries and other organizations.

FUTURE OF MOBILE

Mobile has made a tremendous impact in such a short time, and the future outlook is bright. The expectation is that mobile will continue to dominate the technology landscape. Benedict Evans, partner at Andreessen Horowitz (a venture capital firm in Silicon Valley) who maintains a close watch on the state and future of mobile technology, wrote on his blog in December 2015 about various aspects of mobile that we need to be aware. The blog post is titled "16 Mobile Theses," and it provides insights into the state of mobile technology that is valuable not only for businesses and other organizations but libraries as well. Among the things mentioned in his 16 theses were

- mobile is the new central ecosystem of tech
- mobile is the Internet
- the future of productivity
- apps and the web
- the Internet of Things[12]

These are aspects that might be of particular interest to libraries because they give us further insight into the importance of mobile and how it will shape the future of technology. Thus, it gives us a glimpse into how our users and potential users will use mobile and what their expectations might be.

Mobile Is the New Central Ecosystem of Tech

Evans describes mobile as the new central ecosystem of the tech industry. Each new generation in technology represents a change in scale that makes it the center of innovation and investment in hardware, software, and company creation. He asserts that the mobile ecosystem is potentially headed to 10 times the scale of the PC industry. He describes the smartphone as the sun of this new ecosystem and says that almost everything in the tech industry will orbit around it. While libraries are not solely part of the tech industry, it is important to maintain an awareness of the current technology and how people are using it because those same people are library users, or potential library users, who have a set of expectations for information consumption that the library needs to be able to meet.

Mobile Is the Internet

After smartphones were initially introduced we developed this idea of the mobile Internet versus the desktop Internet. The mobile Internet was great, but it was limited by the small screen of the mobile device that it was being viewed on. Websites built for mobile devices were often stripped-down versions of a full website that provided the most basic information to users. That is no longer the case as we can see by the project mobile Internet subscriptions (Figure 8.1). Evans says that we should stop thinking of the Internet in terms of mobile versus desktop because mobile is the Internet.

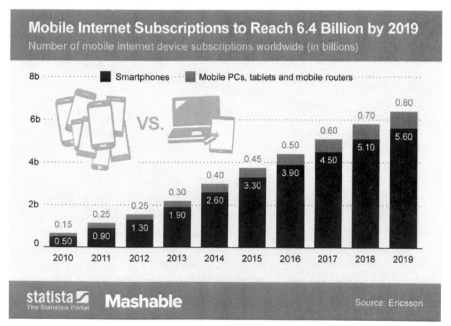

Figure 8.1. Mobile Internet Subscriptions by 2019. *Ericsson / Statista.com*

The capabilities of smartphones make them more sophisticated as an Internet platform than the PC. He goes on to say that it is the PC that has the limited, cut-down version of the Internet. We can easily see the point he is making if we look at how mobile has shaped the Internet in terms of organizational strategy to reach mobile users and how mobile has come to shape web design with the introduction of responsive web design, a mobile-first approach to developing websites.

The Future of Productivity

Mobile is the future of productivity. Mobile devices have long been criticized for lacking certain features that enable them to replace the productivity one could do on a PC. The argument was that there were certain things that you could not do on a mobile device that you could do on a PC with a keyboard and mouse, such as work in Excel and PowerPoint. With Microsoft developing mobile apps for their Office products, that is no longer the case. Software development is changing as more things move to the cloud and can be done using mobile devices. For example, many integrated library systems are moving to the cloud and no longer require the downloading of clients to a specific computer. As long as you have access to a browser, be it on a PC or mobile device, you can access your library system to change circulation policies, update catalog records, or add items to your discovery layer. Mobile devices are becoming more robust so that things we traditionally thought had to be done on a PC can now be done on a mobile device.

Apps and the Web

With the introduction of both the App Store and Google Play, there has been a debate over mobile apps and websites. Each has different unique features. Organizations often debate whether or not they should develop an app for users to download. Evans says that the debate over apps and the web boils down to a simple, less technical question: Do people want to put the organization's icon on their device? This is a point of conversation in libraries as well. How do we provide the best access to our users? Do we do that through responsive web design? An app? Or both? Some libraries simply do not have the technical expertise to build an app, but there are companies now that focus exclusively on developing mobile applications for libraries, such as Boopsie. The answer to the question of should we build an app lies in looking at the usage behavior of your users. That could be done through surveys to see what your users prefer. Some libraries like to have both a mobile-friendly website and a library app. While there are those who say that mobile apps are dead, users are still using a variety of different apps for their personal and professional needs.

Internet of Things

The Internet of Things has been hailed as the next big thing in technology. The Internet of Things revolves around the interconnectivity of tools such as mobile devices, cars, and household appliances to name a few. It is machine-to-machine communication that is built on cloud computing and a vast array of network-connected data-gathering sensors. The Internet of Things allows the network of physical objects to exchange and collect data between devices. In 2014, the Pew Research Center published an article about what the Internet of Things means for libraries. The article states that the Internet of Things is a fourth digital revolution, and it is tied to another revolution that is moving toward gigabyte connectivity. It will reshape who librarians are and what they do. It will lead to the reconfiguration of library space, and it will redefine the role that libraries play in their communities.[13]

The Internet of Things offers some promising opportunities to libraries. An article in OCLC's Next Space lists the following things that make the Internet of Things promising for libraries:

- inventory control
- mobile payments, ticketing, and event registration
- access and authentication
- climate and room configuration
- mobile reference
- resource availability for both content and physical plant
- smart books
- gaming and augmented reality
- object-based learning
- assistive technology[14]

Despite the opportunities of the Internet of Things there are some concerns not only for libraries but for society as whole. One of the biggest concerns is that of privacy, security, and hacking. Currently there are no security protocols in place, and many are calling for these security protocols to be developed. Another concern is the expense involved with participating in these new types of technologies. Regardless of the concerns, it is projected that the Internet of Things will enter into the mainstream by 2020 (Figure 8.2).

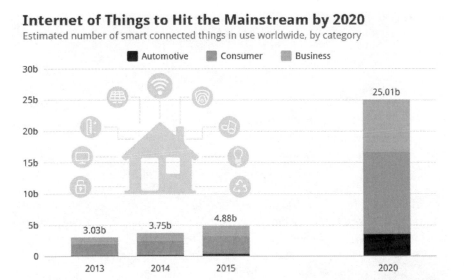

Figure 8.2. Internet of Things into the Mainstream by 2020. *Gartner / Statista.com*

WHAT THE FUTURE MEANS FOR LIBRARIES

The influence of mobile technology has had a profound influence on society. Information behaviors have changed now that mobile technology enables users to be constantly connected to information. Expectations have changed as well. Users expect to be able to access information and complete many of the tasks traditionally done on a desktop on the device of their choosing, which is becoming more and more a mobile device. Organizations, libraries included, have been adjusting to this rapid shift in technology and looking for ways to satisfy and exceed user expectations.

Technology in general has played a big role in the development of libraries over the years. The introduction of the online catalog that enabled users to search a listing of books available in the library and to access online subscription databases and other electronic resources has made information readily available to users in a variety of formats. Technology will continue to shape libraries. TeachThought, a platform that explores learning innovation, stated that the library of the future will contain more technology. In an article titled "10 Ways the Library of the Future Will Be Different," TeachThought states,

> Probably the most obvious direction libraries will trend involves more seamless integration of technologies at a faster, more sophisticated pace than even now.

With so many exciting new gadgets and concepts such as ebook readers, tablet PCs, open source, and more, they [libraries] have of resources on hand to meet community demands. Books, sadly, do not hold the same collective appeal as the shiny new gadgets, but enterprising librarians know that they can still bring literature to the masses by utilizing its lust for technology.[15]

In addition to technology becoming more and more prevalent in libraries, the roles and skills of library staff will continue to evolve and shift to meet the demands of users. In relation to mobile technology, this will mean that library staff will need to maintain an awareness of useful mobile applications, as more of our users are equipped with mobile devices. Libraries are starting to compile lists of useful mobile applications in a variety of different categories ranging from productivity to entertainment to reference and making those lists accessible to users on their websites and LibGuides. Library staff will need to be aware of the different types of mobile devices and the various features and functionalities available on them. With more users having access to mobile devices, they will be seeking assistance with the operation of their devices. At Georgetown College, my library, we have been experiencing this more frequently. Our users have stopped by our research desk seeking assistance with their devices. Providing this assistance is another way that we can showcase our value to our users. Libraries will need to continue to

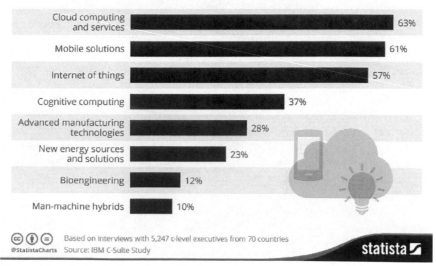

These Technologies Will Shape the Near Future

% of CxOs thinking these technologies will be particularly important in the next 3-5 years

Cloud computing and services — 63%
Mobile solutions — 61%
Internet of things — 57%
Cognitive computing — 37%
Advanced manufacturing technologies — 28%
New energy sources and solutions — 23%
Bioengineering — 12%
Man-machine hybrids — 10%

Based on interviews with 5,247 c-level executives from 70 countries
@StatistaCharts Source: IBM C-Suite Study

statista

Figure 8.3. Technologies That Will Shape the Near Future. *IBM C-Suite Study / Statista.com*

eval￼uate services and resources to determine how they translate to the mobile landscape.

Libraries and librarians are already doing some amazing things when it comes to mobile technology. We have seen libraries checking out mobile devices, developing responsive websites and mobile apps, testing wearable technology like Oculus Rift, and looking at other ways to make mobile technology available to users. Mobile technology will continue to shape the future and libraries (Figure 8.3). Regardless of where the future of mobile takes us, it will continue to offer libraries an opportunity to interact with and reach our users in an exciting and engaging way.

NOTES

1. Android Developers. "Near Field Communication." 2014. http://developer.android.com/guide/topics/connectivity/nfc/index.html

2. HID Global. "HID Global Launches First University Pilot of NFC Smartphones Carrying Digital Keys for Access Control." September 14, 2011. http://www.hidglobal.com/press-releases/hid-global-launches-first-university-pilot-nfc-smart-phones-carrying-digital-keys

3. Nosowitz, Dan. "Use Your Smartphone at the Library, but not to Read Books." *Popular Science*, July 8, 2013. http://www.popsci.com/gadgets/article/2013-07/use-your-smartphone-library-not-read-books

4. Nosowitz, Dan. "Use Your Smartphone at the Library, but Not to Read Books." *Popular Science*, July 8, 2013. http://www.popsci.com/gadgets/article/2013-07/use-your-smartphone-library-not-read-books

5. Velazco, Chris. "MLB's iBeacon Experiment May Signal a Whole New Ball Game for Location Tracking." *TechCrunch*, September 29, 2013. http://techcrunch.com/2013/09/29/mlbs-ibeacon-experiment-may-signal-a-whole-new-ball-game-for-location-tracking/

6. Gorman, Michael. "San Francisco Giants (and Most of MLB) Adopt Apple's iBeacon for an Enhanced Ballpark Experience." Endgadget, March 28, 2014. http://www.engadget.com/2014/03/28/san-francisco-giants-mlb-ibeacon/

7. Capira Technologies. "iBeacon Library App Integration." 2016. http://www.capiratech.com/products/capiramobile/ibeacon/

8. Enis, Matt. "Beacon Technology Developed by Two Library App Makers." *Library Journal*, November 18, 2014. http://lj.libraryjournal.com/2014/11/marketing/beacon-technology-deployed-by-two-library-app-makers/#_

9. Ibid.

10. Google. "Eddystone." https://github.com/google/eddystone

11. Amadeo, Ron. "Meet Google's 'Eddystone'—a Flexible, Open Source iBeacon Fighter." *Ars Technica*, July 14, 2015. http://arstechnica.com/gadgets/2015/07/meet-googles-eddystone-a-flexible-open-source-ibeacon-fighter/

12. Evans, Benedict. "16 Mobile Theses." *Benedict Evans Blog*, December 18, 2015. http://ben-evans.com/benedictevans/2015/12/15/16-mobile-theses

13. Raine, Lee. "The Internet of Things and What It Means for Libraries." Pew Research Center, October 28, 2014. http://www.pewinternet.org/2014/10/28/the-internet-of-things-and-what-it-mean-for-librarians/

14. OCLC. "Libraries and the Internet of Things." Next Space, February 15, 2015. https://www.oclc.org/publications/nextspace/articles/issue24/librariesandthein ternetofthings.en.html

15. TeachThought. "10 Ways The Library gf the Future Will Be Different." November 18, 2012. http://www.teachthought.com/trends/10-ways-the-library-of-the-future-will-be-different/

Index

About the Author

Ben Rawlins is the director of library services at the Ensor Learning Resource Center at Georgetown College in Georgetown, Kentucky. In addition to his role as director, Ben is in charge of the library's website, mobile presence, and other digital services. He has developed several mobile websites and applications for the iOS and Android operating systems. He has presented at several national conferences—including the LITA National Forum, ALA Midwinter, ALA Annual, Computers in Libraries, and Handheld Librarian Conference—on a variety of topics related to mobile services in libraries. He is the author of the book *Mobile Devices: A Practical Guide for Librarians* and has co-authored articles that have been published in *Kentucky Libraries*, *The Reference Librarian*, and *Mobile Library Services: Best Practices*. In addition to his masters in library and information science, Ben also holds an MA in history. He can be reached at brawlins4@gmail.com.